Energy Science, Engineering and Technology

HIGH TEMPERATURE ELECTROLYSIS IN LARGE-SCALE HYDROGEN PRODUCTION

ENERGY SCIENCE, ENGINEERING AND TECHNOLOGY

**Computational Techniques: The Multiphase CFD Approach to
Fluidization and Green Energy Technologies (includes CD-ROM)**
Dimitri Gidaspow and Veeraya Jiradilok (Authors)
2009. 978-1-60876-024-4

Cool Power: Natural Ventilation Systems in Historic Buildings
Carla Balocco and Giuseppe Grazzini (Authors)
2009. 978-1-60876-129-6

**Radial-Bias-Combustion and Central-Fuel-Rich Swirl Pulverized
Coal Burners for Wall-Fired Boilers**
Zhengqi Li (Author)
2009. 978-1-60876-455-6

Bioethanol: Production, Benefits and Economics
Jason B. Erbaum (Editor)
2009. 978-1-61668-000-8

Nuclear Fuels: Manufacturing Processes, Forms and Safety
Antoine Lesage and Jérôme Tondreau (Editors)
2010. 978-1-60876-326-9

Bio Resource-Based Energy for Sustainable Societies
*K.A. Vogt, D.J. Vogt, M. Shelton,
R. Cawston, L. Nackley et al. (Authors)*
2010. 978-1-60876-803-5

Stages of Deployment of Syngas Cleaning Technologies
Filomena Pinto, Rui Neto André and I. Gulyurtlu (Authors)
2010. 978-1-61668-586-7

Photovoltaics: Developments, Applications and Impact
Hideki Tanaka and Kiyoshi Yamashita (Editors)
2010. 978-1-60876-022-0

Thin-Film Solar Cells
Abban Sahin and Hakim Kaya (Editors)
2010. 978-1-60741-818-4

Advanced Materials and Systems for Energy Conversion:
Fundamentals and Applications
Yong X. Gan (Author)
2010. 978-1-60876-349-8

A Solar Car Primer
Eric F. Thacher (Author)
2010. 978-1-60876-161-6

Handbook of Sustainable Energy
W. H. Lee and V. G. Cho (Editors)
2010. 978-1-60876-263-7

Thin-Film Solar Cells
Abban Sahin and Hakim Kaya (Editors)
2010. 978-1-60741-818-4

Advanced Materials and Systems for Energy Conversion:
Fundamentals and Applications
Yong X. Gan (Author)
2010. 978-1-60876-349-8

A Solar Car Primer
Eric F. Thacher (Author)
2010. 978-1-60876-161-6

Handbook of Sustainable Energy
W. H. Lee and V. G. Cho (Editors)
2010. 978-1-60876-263-7

Jatropha Curcas as a Premier Biofuel:
Cost, Growing and Management
Claude Ponterio and Costanza Ferra (Editors)
2010. 978-1-60876-003-9

Ethanol Biofuel Production
Bratt P. Haas (Editor)
2010. 978-1-60876-086-2

Biodiesel Handling and Use Guide
Bryan D. O'Connery (Editor)
2010. 978-1-60876-138-8

Syngas Generation from Hydrocarbons and
Oxygenates with Structured Catalysts
Vladislav Sadykov, L. Bobrova, S. Pavlova, V. Simagina,
L. Makarshin, V. Parmon, Julian R. H. Ross
and Claude Mirodatos (Authors)
2010. 978-1-60876-323-8

Corn Straw and Biomass Blends:
Combustion Characteristics and NO Formation
Zhengqi Li (Author)
2010. 978-1-60876-578-2

Introduction to Power Generation Technologies
Andreas Poullikkas (Author)
2010. 978-1-60876-472-3

CFD Modeling and Analysis of Different Novel
Designs of Air-Breathing Pem Fuel Cells
Maher A.R. Sadiq Al-Baghdadi (Author)
2010. 978-1-60876-489-1

A Sociological Look at Biofuels: Understanding the
Past/Prospects for the Future
Michael S. Carolan (Author)
2010. 978-1-60876-708-3

Direct Methanol Fuel Cells
A. S. Arico, V. Baglio and V. Antonucci (Authors)
2010. 978-1-60876-865-3

Ejectors and their Usefulness in the Energy Savings
Latra Boumaraf, André Lallemand
and Philippe Haberschill (Authors)
2010. 978-1-61668-210-1

**High Temperature Electrolysis in
Large-Scale Hydrogen Production**
Yu Bo and Xu Jingming (Authors)
2010. 978-1-61668-297-2

A Solar Car Primer
Eric F. Thacher (Author)
2010. 978-1-61668-382-5

Biofuels from Fischer-Tropsch Synthesis
M. Ojeda and S. Rojas (Editors)
2010. 978-1-61668-366-5

Transient Diffusion in Nuclear Fuels Processes
Kal Renganathan Sharma (Author)
2010. 978-1-61668-369-6

Coal Combustion Research
Christopher T. Grace (Editor)
2010. 978-1-61668-423-5

**Utilisation and Development of Solar
and Wind Resources**
Abdeen Mustafa Omer (Author)
2010. 978-1-61668-497-6

Shale Gas Development
Katelyn M. Nash (Editor)
2010. 978-1-61668-545-4

Coal Combustion Research
Christopher T. Grace (Editor)
2010. 978-1-61668-646-8

**High Temperature Electrolysis in
Large-Scale Hydrogen Production**
Yu Bo and Xu Jingming (Author)
2010. 978-1-61668-697-0

Shale Gas Development
Katelyn M. Nash (Editor)
2010. 978-1-61668-728-1

Transient Diffusion in Nuclear Fuels Processes
Kal Renganathan Sharma (Author)
2010. 978-1-61668-735-9

Sustainable Resilence of Energy Systems
Naim Hamdia Afgan (Author)
2010. 978-1-61668-738-0

Biofuels from Fischer-Tropsch Synthesis
M. Ojeda and S. Rojas (Editors)
2010. 978-1-61668-820-2

Energy Science, Engineering and Technology

HIGH TEMPERATURE ELECTROLYSIS IN LARGE-SCALE HYDROGEN PRODUCTION

YU BO

AND

XU JINGMING

Nova Science Publishers, Inc.
New York

For permission to use material from this book please contact us:
Telephone 631-231-7269; Fax 631-231-8175
Web Site: http://www.novapublishers.com

NOTICE TO THE READER

The Publisher has taken reasonable care in the preparation of this book, but makes no expressed or implied warranty of any kind and assumes no responsibility for any errors or omissions. No liability is assumed for incidental or consequential damages in connection with or arising out of information contained in this book. The Publisher shall not be liable for any special, consequential, or exemplary damages resulting, in whole or in part, from the readers' use of, or reliance upon, this material.

Independent verification should be sought for any data, advice or recommendations contained in this book. In addition, no responsibility is assumed by the publisher for any injury and/or damage to persons or property arising from any methods, products, instructions, ideas or otherwise contained in this publication.

This publication is designed to provide accurate and authoritative information with regard to the subject matter covered herein. It is sold with the clear understanding that the Publisher is not engaged in rendering legal or any other professional services. If legal or any other expert assistance is required, the services of a competent person should be sought. FROM A DECLARATION OF PARTICIPANTS JOINTLY ADOPTED BY A COMMITTEE OF THE AMERICAN BAR ASSOCIATION AND A COMMITTEE OF PUBLISHERS.

LIBRARY OF CONGRESS CATALOGING-IN-PUBLICATION DATA

Available upon Request
ISBN: 978-1-61668-297-2

Published by Nova Science Publishers, Inc. ✦ New York

CONTENTS

PREFACE

With the rapid development of hydrogen use technology, such as fuel cells, large-scale hydrogen production is becoming more and more of a concern worldwide in recent years. High temperature electrolysis (HTE), which is the highly efficient electrolysis of steam at high temperature and utilizes the heat and electrical power supplied simultaneously by advanced nuclear reactor, provides a very promising way for massive production of hydrogen in the future. Operation at high temperature reduces the electrical energy requirement for the electrolysis and also increases the thermal efficiency of the power-generating cycle. It is expected that, through the combination of a high-temperature reactor and a high temperature electrolysis facility, the process will achieve an overall thermal conversion efficiency of 48~59%. Planar solid oxide electrolysis cell (SOEC) technology is being developed because it has the best potential for high efficiency due to minimized voltage and current losses. Furthermore, Modular design of the stack can increase the reliability and flexibility for different scale of demands.

This chapter provides an overview of HTE technology including its key characteristics and challenges of SOEC development. The chapter also introduces the mechanism and system configuration of hydrogen production through HTE, analyzes the requirements for heat sources integrated with HTE. In addition, the thermo- hydrogen transformation efficiency has been studied in detail by the author based on thermodynamics, reaction kinetics and system principals. The relationship of thermo-hydrogen conversion efficiency with temperature is clearly illustrated.

1. INTRODUCTION

The growing gap between the increasing dependence on conventional fossil fuels and the shrinking supply of primary energy sources has spurred the need to promote clean new energy sources, which becomes now even more urgent due to global climate change and environmental destruction widely recognized worldwide in recent years[1-2]. According to the prediction by the World Energy Council (as shown in figure 1), an average growth rate of about 1.3% per year is expected for global energy consumption up to the year 2050[3]. Such an increase in energy consumption in this period can only be appropriately covered by fossil fuel and nuclear energy carrier despite the growing share of renewable energies. But the associated rise in CO_2 emissions and other types of environmental pollution as well as the increasing shortage of fossil fuel resources, makes the search for suitable alternatives for energy supply indispensable.

Hydrogen, the most abundant element in the earth is as the most promising secondary energy carrier because it is storable, transportable and environmental friendly [4]. A hydrogen based economy and society is now proposed that could make an essential contribution to energy supply and allow expansion of energy production while improving environmental quality as shown in figure 2 [5]. Hydrogen can be used as a fuel for heating, electricity generation (using fuel cells) and vehicles [6-9]. It also can be used as a raw material for many chemical processes, such as ammonia and methanol synthesis, iron ore processing, petroleum processing and others [10-12]. With the rapid growth of energy consumption, large-scale hydrogen production is becoming more and more concerned. Nowadays, the world's annual consumption of H_2 is about 50 million tons, which is used primarily for ammonia production and liquid fuel conversion [13-16]. If the cost reduction goal for fuel-cell vehicles is reached, the transportation sector may ultimately be fueled by H_2. This implies a H_2 consumption growth by one/two orders of magnitude for the next several decades. It is clear that hydrogen will take a key role in the coming future.

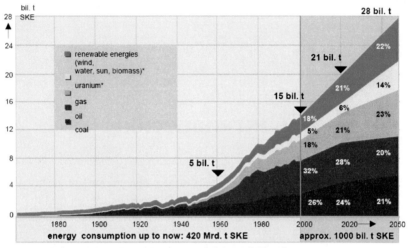

*power production by nuclear and renewable energies assessed by substitution method

Source: WEC, IEA/OECD 2001/2002.

Figure 1. Global energy consumption [3].

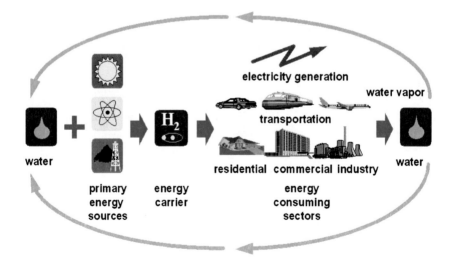

Figure 2. Hydrogen Based Economy and Society [5].

However, this vision of the future relies not only on the development of hydrogen use technology but also on the development of hydrogen production technology. The massive production of hydrogen must be competitive economically and environmental friendly. At present, hydrogen is produced primarily via fossil fuel and electrolysis, about 38% from steam methane reforming (SMR), 30% from coal gasification, 28% from petroleum and 4% from the conventional water electrolysis [17-18]. Seen from a long-term perspective, SMR is not a viable process for more massive hydrogen production since such a conversion process consumes fossil fuel and releases green house gases CO_2 into the air, which largely reduces the benefits of using hydrogen. Conventional water electrolysis is a well-established technique, which is capable of producing highly pure hydrogen without CO_2 direct emissions. However, this process consumes a lot of electricity which results in an unsatisfactory efficiency (about 27%) and leads to high cost [19-20]. Neither of these two conventional methods could meet the requirements for hydrogen economy in the future.

High Temperature Electrolysis (HTE), which is the electrolysis of steam at high temperature, offers a very promising way to produce hydrogen with high efficiency. Compared with conventional water electrolysis, HTE reduces the electrical energy requirement for the electrolysis and increases power cycle efficiency. In addition, high temperature systems can promote the electrode activity and lessen the overpotential [21]. Consequently, it is possible to improve the hydrogen production density through the increase of electric current density and decrease of the polarization losses effectively. Thus HTE is advantageous from both thermodynamic and kinetic standpoints. Moreover, as HTE process is the reverse concept of the solid oxide fuel cell (SOFC) technology, all experiences obtained from R&D on SOFC will be also applicable to the R&D on HTE process [22-24].

The challenge in the coming years, driven by the perspective of the transition to the hydrogen economy, is to turn the ideas and the promises of past decades into industrial reality [25]. This means HTE process has now to prove viability and feasibility leading to the near-term steps of demonstration. The first priority should be intensive efforts to acquire reliable data in the fields of thermodynamics and kinetics of HTE process, material behavior, and SOEC electrochemical performance to support process simulation and technological optimization. Efforts must also be directed at evaluation and optimization of targets, which will in turn drive innovation, technological improvement and breakthroughs. So the objective of this paper is to review the state of the art of HTE technology, to present the fundamental characteristics of HTE process, to discuss emerging options, to address difficulties and challenges, and to define the scope for future improvement.

2. PRINCIPAL CONSIDERATIONS FOR HTE

2.1. BASIC PRINCIPLE AND MAIN COMPONENTS

The realization of HTE needs a reliable and feasible system, which is one of the most difficulties to electrolyze water at high temperature. HTE process using SOEC is a reverse reaction of SOFC developed vigorously in the world. A single SOEC consists of three different layers as shown in figure 3 [26-27]. The middle layer (gray) is oxide ion-conducting electrolyte that is gastight and electron insulating. The lower layer is anode, air or O_2 electrode. The upper layer is cathode made of YSZ and Ni. The electrode must have a good electron and oxide ion conductivity and a porous structure to get a large three phase boundary area where the gases are easy to pass through and the three species (gas molecule, oxygen ions and electrons) can meet and react. In addition, sealing and interconnecting materials should be utilized in planar designed SOEC stacks.

A cell voltage is established between the electrodes when gases with different oxygen partial pressures are fed into the electrodes as given by the Nernst equation. The basic principle of HTE is shown in figure 4 [28-29] and can be described as follows.

Cathode Reaction: $H_2O + 2e \rightarrow H_2 + O^{2-}$

Anode Reaction: $O^{2-} \rightarrow 2e + \frac{1}{2}O_2$

Overall Reaction: $H_2O \rightarrow H_2 + \frac{1}{2}O_2$

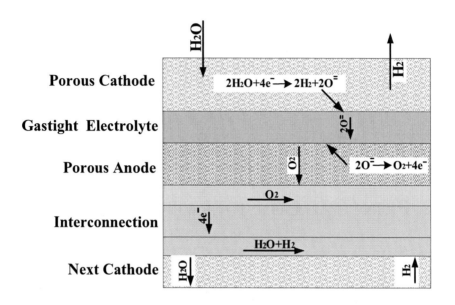

Figure 3. Main components of solid oxide electrolytic cell (SOEC) [26].

In a SOEC mode, a mixture of steam and hydrogen to maintain reducing conditions at 750 to 950℃ is introduced into the high temperature cathode. Then it penetrates through the porous cathode and arrives at the interface between the cathode and the solid oxide electrolyte, where a water molecule is electrically dissociated into a hydrogen and an oxygen anion by two electrons transported from the anode through externally provided electricity. The produced hydrogen is back-diffused to the cathode chamber. Simultaneously oxygen ions are drawn through the electrolyte by the electrochemical potential liberating their electrons and recombining to form oxygen molecules on the surface of anode side [29, 30]. The cathode is graded with a nickel-zirconia cermet layer immediately adjacent to the electrolyte and a pure nickel outer layer. The partially oxidized nickel element of $Ni-ZrO_2$ cathode must be reduced before conducting an electrolytic chemical reduction of steam. Besides the reduction environment for keeping the pure nickel metal is required continuously during operation and shut-down by introducing hydrogen inside the cathode chamber. The entering steam-hydrogen mixture maybe contain as much as 90 vol.% steam, while the exiting mixture gas maybe contain as much as 90 vol.% H_2. The existing water and hydrogen gas mixture passes through a condensing or membrane separator to purify the hydrogen [31].

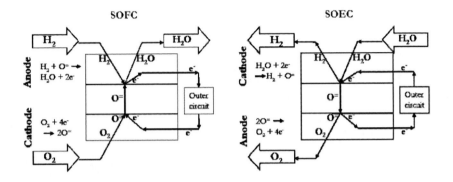

Figure 4. Reaction Mechanisms of SOFC and SOEC [28].

2.2. EFFICIENCY ANALYSIS OF HTE

Because HTE of steam needs electricity and heat, the net thermal-to-hydrogen energy efficiency of the process is governed critically by the operating temperature and the power cycle efficiency [32-34]. In a HTE process, the total heat Q_N for the electrolysis comes from two parts. One is the heat $Q_{N,el}$ for power conversion system and the another is the heat $Q_{N,es}$ directly for HTE process, that is:

$$Q_N = Q_{N,el} + Q_{N,es}$$

In the efficiency analysis, it is assumed that the coolant outlet, the power turbine inlet and the HTE operation temperatures are equal because change of operation temperature from inlet to the outlet of the HTE modules is negligible due to a series of HTE cell modules in one hydrogen plant.

$\eta_{H,HTSE,ideal}$, the ideal value of the overall thermal-to-hydrogen production efficiency from HTE, is the ratio of the amount of energy that is carried by unit amount of hydrogen produced, $Q_{H,out}$ to the total thermal energy Q_N required in the steam electrolysis process to produce the unit amount of hydrogen [32].

$$\eta_{H,HTSE,ideal} = \frac{Q_{H,out}}{Q_N} = \frac{\Delta H}{Q_{N,el} + Q_{N,es}}$$

2.2.1. Thermodynamic Efficiency

If the electrical energy utilized in the electrolysis is obtained directly from a power plant, e.g. HTGR power plant. For an electrolytic cell, the minimum electrical work necessary for the process is the electrochemical potential. For a reversible and steady state, constant (T, P)

$$W_{min} = -\Delta G$$

Therefore, $Q_{N,el}$ can be expressed as:

$$Q_{N,el} = \frac{\Delta G}{\eta_{el}}$$

where η_{el} is the net power cycle efficiency. Therefore, it can be deduced that:

$$\eta_{H,HTSE,ideal} = \frac{\Delta H}{\dfrac{\Delta G}{\eta_{el}} + Q_{N,es}} = \eta_{el} \frac{\Delta H}{\Delta H - (1 - \eta_{el})Q_{N,es}} = \frac{\Delta H}{\Delta H + \dfrac{1 - \eta_{el}}{\eta_{el}} \cdot \Delta G}$$

As the thermodynamic parameters (ΔH, ΔG, and ΔS) are the function of the temperature, we calculated the thermodynamic parameters at various temperatures according to Kirchhoff's equation, entropy equation and the relation between ΔG and the equilibrium potential shown as Eq. (1), (2) and (3). Table 1 shows the thermodynamic parameters at 298.15K and 101.325KPa. The thermodynamic parameters at the range of 298.15~ 1473.15K were calculated and shown in figure 5.

$$\Delta H(t) = \Delta H^{o}_{298.15} + \int_{298.15}^{T} \Delta C_p dT \tag{1}$$

$$\Delta S(t) = S^{o}_{298.15} + \int_{298.15}^{T} \Delta C_p / T dT \tag{2}$$

$$E(t) = -\frac{\Delta G(t)}{nF} \tag{3}$$

Table 1. Thermodynamic parameters for
HTE at 298.15K and 101.325KPa

	ΔG (KJ/mol)	ΔH (KJ/mol)	S(KJ/ mol)	$\Delta VapH$ (KJ/mol)	Heat Capacity(Cp) Jmol-1K-1
H2(g)	0	0	0.131	-	29.07-0.836×10-3T+20.1×10-7T2
O2(g)	0	0	0.205	-	25.72+12.98×10-3T-38.6×10-7T2
H2O(l)	237.2	285.8	0.07	40.7	75.30
H2O(g)	-	-	-	-	30.36+9.61×10-3T+11.8×10-7T2

(g) and (l) refer to gas phase and liquid phase respectively.

From figure 5, it could be found that ΔH increases slightly with the increasing temperature. ΔG decreases with the temperature, due to increased Q. The ratio of ΔG to ΔH is about 93% at 373.15K and about 70% at 1273.15K.

As the electrolysis temperature increases, ΔG decreases so that the $\eta_{H,HTSE,ideal}$ will increase. Also, It can be seen that $\eta_{H,HTSE,ideal} = \eta_{el}$ under the condition of conventional electrolysis because of no or very limited heat input. Consequently, thermal hydrogen conversion efficiency of HTE must exceed that of conventional electrolysis and increases with the increasing operation temperature.

As above mentioned, hydrogen production efficiency of HTE strongly depends on operation temperature and power cycle efficiency. Temperature sensitivity coefficient (TSC) is defined as the variation of hydrogen production efficiency with temperature. Figure 6 shows the calculation results of TSC vs. power cycle efficiencies. It could be found that the TSC reached at first to the maximum and then almost kept constant without much fluctuation. TSC reached maximum when power cycle efficiency was about 40~50% which indicated that the increase of operation temperature within a certain range will increase the overall efficiency to a large degree and improve the performance of the whole system. But in the range of 40-50% power cycle efficiency, a little operation temperature fluctuation has only limited effect on overall efficiency of hydrogen production. According to this aspect, HTGR is an ideal primary heat source because its power cycle efficiency is just around 40~50% and its outlet temperature can be up to 950℃.

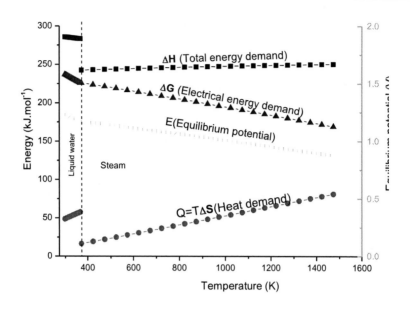

Figure 5. Energy demand for water and steam electrolysis.

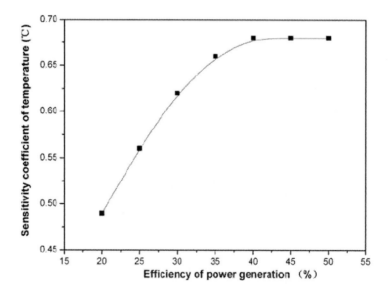

Figure 6. Temperature-sensitivity coefficients under various $\eta_{el.}$

The HTGR coupled with the HTE system can ensure to realize high temperature steam electrolysis process due to the higher outlet temperature of HTGR (up to 950°C), thus, the process is expected to achieve very high thermal conversion efficiency. Currently, high-temperature gas-cooled test reactors have been built and run in China and Japan. In South Africa, China and the United States, there are development plans for the construction of high temperature gas cooled reactor nuclear power plants. Very high temperature gas cooled reactor (VHTR) is one of the proposed six types advanced nuclear reactors called as Generation IV, with the aim of coolant outlet temperature to reach 1000°C.

In addition to operating temperature, the power cycle efficiency (η_{el}), electrolysis efficiency (η_{es}) and thermal efficiency (η_{th}) also have significant influence on $\eta_{overall}$ of the HTE system. The η_{es} refers to the overall efficiency of the SOEC system, including electrochemical efficiency ($\eta_{electrochem}$), faradaic efficiency ($\eta_{Faraday}$) and system efficiency (η_{system}). There is following relation:

$$\eta_{electrochem} = V_{op}(i,\ T)/E(T)$$

Where $V_{op}(i,\ T)$ is approximately proportional to current density i at elevated temperature, where $V_{op}(i,\ T)$ is the operating voltage of the SOEC system at a given current density i and operating temperature T, and $E(T)$ is Nernst potential of the SOEC system at temperature T.

η_{system} is resulted from the parasitic energy consumption of the SOEC system, such as resistance of pipeline, pumping work, AC-DC conversion, etc.

The η_{es} is given by

$$\eta_{es} = \eta_{eletrochem}\, \eta_{Faraday}\, \eta_{system}$$

The η_{th} refers to the thermal utilization efficiency of the HTE system. It considers the thermal exchange efficiency between the HTGR and the SOEC system, the heat dissipation of the HTE system, the heat consumption for preheating excess steam, hydrogen at the cathode, and oxygen at the anode, as well as the waste heat recycling. The η_{th} is given by

$$\eta_{th} = \frac{Q_{SOEC}}{Q_{HTGR}} = \frac{T_{inlet} - T_{outlet}}{T_{inlet}}$$

Where Q_{HTGR} is the thermal energy from HTGR, Q_{SOEC} is the consumed thermal energy in the SOEC system, T_{inlet} is the outlet temperature of HTGR, and T_{outlet} is the outlet temperature of the SOEC system.

The electrical energy consumed in the electrolysis process can be expressed as

$$\Delta G = \eta_{el} \times Q_{el}$$

Where Q_{el} is the consumed thermal energy from HTGR used for generating electricity. Then, the total thermal energy ($Q_{overall}$) required in the electrolysis process from HTGR can be expressed as

$$Q_{overall} = Q_{el} + Q_{th} = 2F[V_{op}(i, T)/\eta_{el} + (V_{th}(T)\text{-}V_{op}(i, T))/\eta_{th}]$$

Where Q_{el} and $2FV_{op}(i, T)/\eta_{el}$ is the consumed thermal energy for generating electricity, and Q_{th} and $2F(V_{th}(T)\text{-}V_{op}(i, T))/\eta_{th}$ is the thermal energy demand in the electrolysis process.

The $\eta_{overall}$ of the HTE system can be defined as the ratio of the energy carried by unit amount of produced hydrogen ($Q_{H, out}$), in terms of a high heat value of hydrogen (HHV=285.8kJ/mol), to the $Q_{overall}$ in the steam electrolysis process [20]. Therefore, the $\eta_{overall}$ can be expressed as

$$\eta_{overall} = \cfrac{HHV}{\cfrac{\Delta G(T)}{\eta_{el}\eta_{es}} + \cfrac{Q_{th}(T)}{\eta_{th}} - \cfrac{\Delta G(T)}{\eta_{es}\eta_{th}}(1 - \eta_{es})}$$

Where,

$$\frac{\Delta G(T)}{\eta_{el}\eta_{es}}$$

is the actual thermal energy consumed for electricity generation.

$$\frac{Q_{th}(T)}{\eta_{th}} - \frac{\Delta G(T)}{\eta_{es}\eta_{th}}(1-\eta_{es})$$

is the actual thermal energy demand in the electrolysis process,

where

$$\frac{\Delta G(T)}{\eta_{es}\eta_{th}}(1-\eta_{es})$$

is the thermal energy converted from polarization and ohmic losses during the electrolysis process.

We can see that $\eta_{overall}$ is examined as a function of individual efficiencies of η_{el}, η_{es}, and η_{th}, which cover almost all of the energy losses in the actual HTE for hydrogen process.

The total differential equation of the $\eta_{overall}$ is written as:

$$d\eta_{overall} = \frac{\partial \eta_{overall}}{\partial \eta_{el}}d\eta_{el} + \frac{\partial \eta_{overall}}{\partial \eta_{es}}d\eta_{es} + \frac{\partial \eta_{overall}}{\partial \eta_{th}}d\eta_{th}$$

Where,

$$\frac{\partial \eta_{overal}}{\partial \eta_{el}} = \frac{HHV \times \Delta G(T)/\eta_{es}}{\left[\dfrac{\Delta G(T)}{\eta_{el}\eta_{es}} + \dfrac{Q_{th}(T)}{\eta_{th}} - \dfrac{\Delta G(T)}{\eta_{es}\eta_{th}}(1-\eta_{es})\right]^2 \eta_{el}^2}$$

$$\frac{\partial \eta_{overal}}{\partial \eta_{es}} = \frac{HHV \times \Delta G(T)(\dfrac{1}{\eta_{el}} - \dfrac{1}{\eta_{th}})}{\left[\dfrac{\Delta G(T)}{\eta_{el}\eta_{es}} + \dfrac{Q_{th}(T)}{\eta_{th}} - \dfrac{\Delta G(T)}{\eta_{es}\eta_{th}}(1-\eta_{es})\right]^2 \eta_{es}^2}$$

$$\frac{\partial \eta_{overal}}{\partial \eta_{th}} = \frac{HHV \times [Q_{th} - \frac{\Delta G(T)}{\eta_{es}}(1-\eta_{es})]}{\left[\frac{\Delta G(T)}{\eta_{el}\eta_{es}} + \frac{Q_{th}(T)}{\eta_{th}} - \frac{\Delta G(T)}{\eta_{es}\eta_{th}}(1-\eta_{es})\right] \eta_{th}^2}$$

In order to analyze the effects of η_{el}, η_{es}, and η_{th} on $\eta_{overall}$, the range of each variable parameter was assumed as follows:

(1) η_{el}: 40~52%.
(2) η_{es}: 60~100%.
(3) η_{th}: 30~90%.
(4) T: 500~1000℃.

2.2.1.1. Power Cycle Efficiency (η_{el})

Figure 7 shows our calculation results of the effect of η_{el} on overall efficiency $\eta_{overall}$. As shown in figure 7, the η_{el} has a significant effect on the $\eta_{overall}$, for the values of $\partial \eta_{overall} / \partial \eta_{el}$ are in the range of 1.175 and 1.062, which means the $\eta_{overall}$ would change more than one unit if the η_{el} changes one unit.

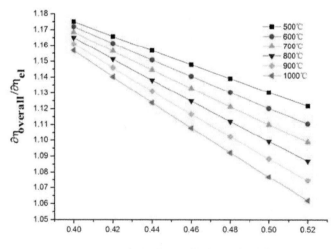

Electricity efficiency (η_{el})

Figure 7. Effect of η_{el} on $\eta_{overall}$ (η_{es}: 100%, η_{th}: 90% assumed).

2.2.1.2. Electrolysis Efficiency (η_{es})

As shown in figure 8, the effect of η_{es} on $\eta_{overall}$ is not as observable as η_{el}. The values of $\partial\eta_{overall}/\partial\eta_{es}$ are only in the range of 0.426 and 0.251. With the increase of η_{es} from 60% to 100%, the effects of η_{es} on $\eta_{overall}$ decrease at various temperatures, and the trends are non-linear. On the other hand, as temperatures increased from 500 ℃ to 1000 ℃, the effect of η_{es} on $\eta_{overall}$ almost stayed constant, as shown in figure 8.

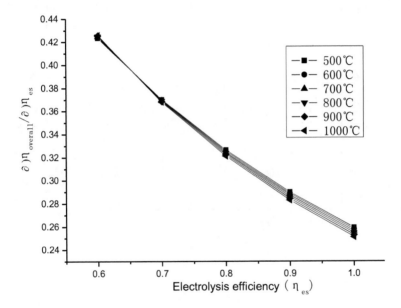

Figure 8. Effect of η_{es} on $\eta_{overall}$. (η_{el}: 45%, η_{th}: 90% assumed).

2.2.1.3. Thermal Efficiency (η_{th})

Figure 9 shows the effect of η_{th} on $\eta_{overall}$. With the increase of η_{th} from 30% to 90%, the effects of η_{th} on $\eta_{overall}$ diminish rapidly at various temperatures, as shown in figure 9. The changing trends show the exponential relationship, which reveals that the values of $\partial\eta_{overall}/\partial\eta_{th}$ change rapidly at low η_{th} (e.g. below 50%), and change slowly with the increase of η_{th}. The values of $\partial\eta_{overall}/\partial\eta_{th}$

dramatically decreased from 0.552 to 0.055 when η_{th} escalated from 30% to 90%. During practical operating processes, the η_{th} of the HTE system is usually above

50%. Therefore, η_{th} values from 50% to 90%, which corresponds to $\partial \eta_{overall} / \partial \eta_{th}$ values between 0.272-0.055, do not have as much effect on $\eta_{overall}$ as compared to η_{el} and η_{es}.

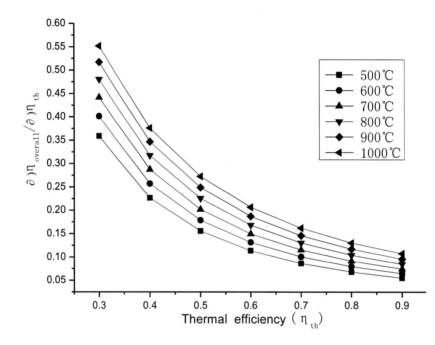

Figure 9. Effect of η_{th} on $\eta_{overall}$. (η_{el}: 45%, η_{es}: 100% assumed).

Consequently, the qualitative effects of electrical, electrolysis, and thermal efficiency on $\eta_{overall}$ can be ordered as $\eta_{el} > \eta_{es} > \eta_{th}$, which is consistent with the conclusion by Shin [35].

In addition, as temperatures increased from 500°C to 1000°C, the effect of η_{th} on $\eta_{overall}$ increased gradually, as shown in figure 9. This is in contrast to what was seen for η_{el} (figure 7). This demonstrates that the effects of η_{th} on $\eta_{overall}$ diminish with increasing η_{th} and decreasing operating temperatures.

2.2.1.4. Quantitative Analysis of the η_{el}, η_{es} and η_{th} Effects

Tables 2, 3, and 4 show the $\partial\eta_{overall}/\partial\eta_{efficiency}$ values of different η_{el}, η_{es}, η_{th}, at various temperatures separately. For comparing the effect of η_{el}, η_{es} and η_{th}, the average values of $\partial\eta_{overall}/\partial\eta_{efficiency}$ at different η_{el}, η_{es}, η_{th} are calculated and normalized, the obtained results are shown in table 5 and figure 10. The results illustrate the quantitative effects of η_{el}, η_{es} and η_{th}. With the increase of temperature, the η_{el} effects decrease slightly, and the η_{th} effects increase slightly, while the η_{es} effects stay almost constant. The quantitative effects of the η_{el}, η_{th}, and η_{th} are about 70%, 22%, and 8%, respectively.

Table 2. $\partial\eta_{overall}/\partial\eta_{el}$ **values of different η_{el} at various temperatures**

Temperature (°C)	ηel						
	0.4	0.42	0.44	0.46	0.48	0.50	0.52
500	1.175	1.1658	1.1568	1.1479	1.1391	1.1304	1.1218
600	1.172	1.1613	1.1508	1.1405	1.1304	1.1203	1.1105
700	1.168	1.1565	1.1446	1.1329	1.1214	1.11	1.0988
800	1.165	1.1514	1.138	1.1249	1.112	1.0993	1.0869
900	1.1611	1.1459	1.1311	1.1166	1.1023	1.0883	1.0746
1000	1.1568	1.1402	1.1238	1.1078	1.0922	1.0769	1.0619

Table 3. $\partial\eta_{overall}/\partial\eta_{es}$ **values of different η_{es} at various temperatures**

Temperature (°C)	ηes				
	0.6	0.7	0.8	0.9	1
500	0.4234	0.3702	0.3264	0.2899	0.2593
600	0.4239	0.3699	0.3255	0.2887	0.2578
700	0.4245	0.3695	0.3246	0.2874	0.2562
800	0.4250	0.3691	0.3236	0.2860	0.2546
900	0.4255	0.3687	0.3225	0.2845	0.2529
1000	0.4259	0.3681	0.3213	0.2830	0.2511

Table 4. $\partial\eta_{overall}/\partial\eta_{th}$ **values of different η_{th} at various temperatures**

Temperature (°C)	ηth				
	0.5	0.6	0.7	0.8	0.9
500	0.1556	0.1136	0.0865	0.0681	0.055

Yu Bo and Xu Jingming

600	0.1787	0.1314	0.1007	0.0796	0.0645
700	0.202	0.1496	0.1153	0.0915	0.0744
800	0.2254	0.1682	0.1304	0.104	0.0848
900	0.2489	0.1872	0.1459	0.1169	0.0958
1000	0.2725	0.2066	0.162	0.1304	0.1072

Table 5. The average values of $\partial\eta_{overall} \big/ \partial\eta_{efficiency}$ at various temperatures

Temperature (°C)	Normalized average values of $\partial\eta_{overall} \big/ \partial\eta_{efficiency}$		
	$\partial\eta_{overall} \big/ \partial\eta_{el}$	$\partial\eta_{overall} \big/ \partial\eta_{es}$	$\partial\eta_{overall} \big/ \partial\eta_{th}$
500	72.771	21.160	6.070
600	71.978	21.020	7.002
700	71.173	20.879	7.948
800	70.353	20.734	8.912
900	69.521	20.588	9.891
1000	68.676	20.437	10.888

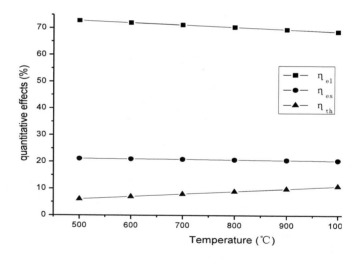

Figure 10. Quantitative effects of η_{el}, η_{es} and η_{th} on $\eta_{overall}$ at various temperatures.

2.2.1.5. Calculations of the $\eta_{overall}$

In this paper, the actual power cycle efficiencies of HTGR at different temperatures were assumed as shown in Table 6 [36]. Table 7 shows the calculated results of the overall system efficiencies under various conditions. As shown in table 7, with the increase of η_{el}, η_{es} and η_{th}, the overall system efficiencies are anticipated to be from 33% to 59%, and the maximum $\eta_{overall}$ can reach 59% at 1000°C. Compared with the $\eta_{overall}$ of well developed conventional alkaline water electrolysis, which is about 27%, the efficiency of the HTGR coupled with the HTE system may be much higher than that of the conventional alkaline water electrolysis. HTE system can be considered as a promising way for highly efficient large-scale hydrogen production.

Table 6. Electrical efficiencies of HTGR at different temperatures

Temperature (°C)	500	600	700	800	900	1000
η_{el}	40%	42%	45%	47%	50%	52%

Table 7. Overall system efficiencies under various conditions

η_{th}	50%	60%	70%	80%	90%
η_{es}	$\eta_{overall}$	$\eta_{overall}$	$\eta_{overall}$	$\eta_{overall}$	$\eta_{overall}$
T=500°C (η_{el}=40%)					
60%	0.3704	0.3530	0.3415	0.3334	0.3273
80%	0.3961	0.3935	0.3917	0.3903	0.3892
100%	0.4132	0.4226	0.4295	0.4348	0.4391
T=600°C (ηel=42%)					
60%	0.3971	0.3790	0.3671	0.3586	0.3523
80%	0.4186	0.4174	0.4166	0.4160	0.4155
100%	0.4327	0.4445	0.4533	0.4602	0.4656
T=700°C (ηel=45%)					
60%	0.4360	0.4165	0.4037	0.3945	0.3877
80%	0.4504	0.4510	0.4515	0.4518	0.4520
100%	0.4595	0.4746	0.4860	0.4949	0.5020
T=800°C (ηel=48%)					
60%	0.4746	0.4544	0.4409	0.4313	0.4242
80%	0.4807	0.4837	0.4859	0.4875	0.4888
100%	0.4845	0.5032	0.5175	0.5288	0.5379
T=900°C (ηel=50%)					
60%	0.5000	0.4805	0.4675	0.4582	0.4513
80%	0.5000	0.5057	0.5099	0.5130	0.5155
100%	0.5000	0.5221	0.5392	0.5527	0.5637
T=1000°C (ηel=52%)					
60%	0.5243	0.5062	0.4941	0.4854	0.4788
80%	0.5180	0.5268	0.5333	0.5383	0.5422
100%	0.5143	0.5400	0.5600	0.5760	0.5890

Based on the above one-dimensional analysis, the efficiency of the HTE system integrated with HTGR further is studied via a two-dimensional simulation method, which changes two parameters simultaneously in a reasonable range while keeping one parameter constant. Compared with one-dimensional analysis method, the effects on overall efficiency were investigated more objectively and accurately. Moreover, the critical concepts of η_{es} and η_{th} were put forward originally, which were very important to determine the optimum electrolysis voltages and operation temperatures in the actual HTE processes. Based on the calculation, the critical value of η_{es} was $\Delta G(T)/\Delta H(T)$. Therefore, in the actual operation, η_{es} should be higher than $\Delta G(T)/\Delta H(T)$ in order to maintain the high hydrogen production efficiency of HTE system. Also, it was very interesting to find that the critical η_{es} was the theoretical maximum efficiency in SOFC mode. Furthermore, the critical value of η_{th} was equal to the value of η_{el}, which means the overall efficiency decreases with the increasing η_{es} if the η_{th} in the actual HTE process is less than the critical value of η_{th}. Therefore, it is very important to control the η_{th} higher than the critical value in the actual HTE process to get high overall system efficiency.

2.2.2. Overpotential Energy Loss

Practically, there will be energy losses that we denote as Q_{loss} which account for the electrolytic overpotential in actual electrolysis processes [37-40]. Therefore, the actual amount of energy required for producing a unit amount of hydrogen via HTE should take into account the relevant loss which should be supplied as heat initially. So $\eta_{H,HTE}$, the actual thermal-to-hydrogen energy production efficiency of the HTE process that we use in the evaluation takes the form:

$$\eta_{H,HTE} = \frac{\Delta H}{\dfrac{\Delta G}{\eta_{el}} + Q_{N,es} + Q_{Loss}}$$

In actual operation, there is deviation from equilibrium with finite rate of electrochemical process.

$\eta = \eta_a + \eta_c + \eta_\Omega$ is used to the *overpotential energy loss*

In general , the overpotential η comes from reaction activation polarization η_a, species concentration polarization η_c and ohmic resistance of anode, electrolyte and cathode η_Ω, as shown in figure 11.

In SOFC mode, the operating voltage of a cell is always less than the theoretical maximum whereas the operating voltage is always higher than the theoretical voltage of SOEC due to the existence of overpotential.

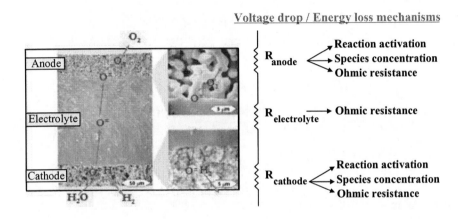

Figure 11. Energy loss mechanism of overpotential in SOEC.

Activation overpotential η_a, results from limiting reaction rates at which one or more electrode reaction steps operate very near the electrode surface. Overall activation overpotential represents that of the rate limiting (slowest) reactions in the series. Relationship between cell current i, and activation overpotential η_a (with no mass transfer effect) is listed as below:

$$i = i_0 \left(\frac{O^{surface}}{O^{bulk}} \exp\left[\frac{-an_e F \eta_a}{RT} \right] - \frac{R^{surface}}{R^{bulk}} \exp\left[\frac{1-a\ n_e F \eta_a}{RT} \right] \right)$$

$$\alpha \approx 0.5; \ O^{surface} = O^{bulk} \ and \ R^{surface} = R^{bulk}$$

$$\Rightarrow i = 2i_0 \sinh\left[\frac{0.5n_e F \eta_a}{RT}\right]$$

Thus

$$\eta_a = \frac{RT}{0.5n_e F} \sinh^{-1}\left(\frac{i}{2i_0}\right)$$

Where "i_0=exchange current density" is specific for each SOEC electrode, and can be found in literature.

The concentration (mass transfer) overpotential η_c , results from the decrease in the reactant concentration and the increase in the product concentration near the electrode surface in comparison to bulk concentrations.

$$i = i_0 \left(\frac{O^{surface}}{O^{bulk}} \exp\left[\frac{-an_e F \eta}{RT}\right] - \frac{R^{surface}}{R^{bulk}} \exp\left[\frac{1-a\ n_e F \eta}{RT}\right] \right)$$

$$\alpha \approx 0.5; \ O^{surface} \neq O^{bulk} \text{ and } R^{surface} \neq R^{bulk}$$

η is a combination of η_a and η_c .

Reduces to:

$$\Rightarrow i = i_0 \left(\left(1 - \frac{i}{i_{L,C}}\right) \exp\left[\frac{-0.5n_e F \eta}{RT}\right] - \left(1 - \frac{i}{i_{L,A}}\right) \exp\left[\frac{0.5n_e F \eta}{RT}\right] \right)$$

Find $\eta = f\ i$ iteratively

$$\Rightarrow \eta_c = \eta - \eta_a$$

Where "$i_{L,C}$=limiting cathodic current" and "$i_{L,A}$=limiting anodic current" are specific for each SOEC electrode, and can be found in literature, or using analytical expression that depends on diffusivity and concentration of species in relevant electrode.

Ohmic (resistance) overpotential η_Ω, is due to resistance of SOEC medium to electron and ion transfer, and is equal to the product of cell resistance and current.

$$R = \sum_j R_j, \quad j = \text{wire, anode, electrolyte, cathode}$$

$$\eta_\Omega = \sum_j R_j \times I$$

$$R_i = 1/\sigma_i * l_i/A_i$$

Where conductivity σ is specific for each SOEC electrode, and can be found from literatures.

Actually, it is difficult to obtain the quantitative values of each overpotential loss under the operation condition. The total overpotential ΔE can be calculated by the measured electrolysis potential (E_m) minus the theoretical electrolysis potential(E_{thor}) at the same conditions, as follows:

$$\Delta E = E_m - E_{thor}$$

Where , the theoretical electrolysis potential can be calculated by thermodynamical data. So the ΔE can be used to discuss the overall effects of these factors. The study conducted by Idaho national laboratory of DOE in USA showed that current density and elevated temperature both had effects on the overall overpotential. The voltage decreased from 1.21V to 1.03V when the temperature increased from 750 to 850℃ at the current density of 0.2A/cm^2.

The research results showed that the ΔE decreased about 70% from 0.45V to 0.13V when the temperature increased from 800 to 900℃ at the current density of 0.2A/cm^2 and the steam content of 75% (see figure 12), which indicated that high temperature is beneficial for the decrease of the cell overpotential and the improvement of the cell kinetics greatly.

From the above analysis, it can be clearly concluded that: the overall thermal-to-hydrogen energy efficiency of HTE increases with increasing operating temperatures due to the following reasons:

- As for the thermodynamic point, high temperature leads to increased direct heat requirement and decreased electrical energy demand
- As for the kinetic point, high temperature helps to promote electrode activity and lower cell overvoltage as well as energy losses

The high temperature gas-cooled reactor is an ideal primary heat source. One of the most important applications of HTE is supposed to contribute a lot to the sustainable development of nuclear energy in the future [41-42]. The combined hydrogen, heat and power generation will give further impetus to higher energy conversion efficiency. Although the development effort for hydrogen production through high temperature steam electrolysis has been substantial currently, the engineering and optimization of SOEC are still at an early stage. Big challenges lies in the development of solid oxide materials for SOEC with high performance especially suitable for operation under high temperature and system integration of SOEC with high temperature heat sources. e.g. HTGR. In addition, several engineering issues must be addressed as part of the development of cell/module for a HTE demonstration.

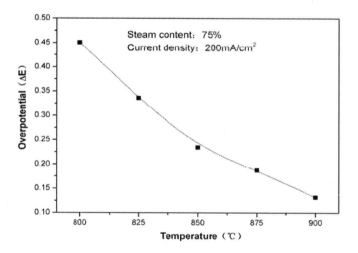

Figure 12. Effect of temperature on overpotential at the current density of $0.2A/cm^2$ and the steam content of 75%.

3. CHALLENGE OF HTE TECHNOLOGY DEVELOPMENT

3.1. SOEC DEVELOPMENT

Operation of a planar solid oxide cells in SOEC mode is fundamentally different from operation in SOFC mode for several reasons, aside from the obvious change in direction of the electrochemical reaction. From the standpoint of heat transfer, operation in SOFC mode typically necessitates the use of significant excess airflow to prevent overheating of the stack due to exothermic hydrogen oxidation reaction and the ohmic heating. However, In the SOEC mode, the steam reduction reaction is endothermic, so the net heat generation may be negative at low current densities, increasing to zero at the thermal-neutral voltage and becoming positive at higher current densities. The thermal-neutral voltage can be predicted from direct using the first law of thermodynamics to the overall system and increase only slightly in magnitude over the typical operating temperature range for SOEC from 1.287V at 800°C to 1.292V at 1000°C which yields current densities in the 0.2 to 0.4A/cm^2. The specific materials for electrodes and electrolyte and the geometry of unit cell can change according to the operating temperature in order to provide an optimized performance. However, the size of an individual cell should be limited to about 20×20 cm^2 because of the shrinkage in a sintering process.

The challenges for SOEC development are mainly focused on the performance and optimization of materials as well as the proper operating conditions which are listed as following [43-44]:

- High temperature environment

- Electrode and electrolyte electronic and ionic conductivity
- Stability and durability of electrode and electrolyte materials
- Thermomechanical matching of the solid oxide electrode and electrolyte materials
- Temperature is a conflicting effect on performance, durability and cost which indicate we should take consideration of a variety of cell materials

3.1.1. Cell Sealing

Sealing is a very important concern for solid oxide stacks used in a HTE mode to generate hydrogen. Because hydrogen is the primary valuable product for electrolysis and due to hydrogen's small size, the gas molecules can quickly leak out of high temperature collection ducts, which will negatively affect the hydrogen gas purity and efficiency. Figure 13 shows schematic of seals typically found in a SOFC stack with metallic internal gas manifolds and a metallic bipolar plate [45]. Common seals include: (a) cell to metal frame; (b) metal frame to metal interconnect; (c) frame/interconnect pair to electrically insulating spacer; (d) stack to base manifold plate, and (e) cell electrode edge or electrolyte to interconnect edge. In many instances, an edge seal will be between zirconia electrolyte and a high-temperature metal frame (or housing). The sealant must endure both the oxidizing environment of the anode area and the reducing condition of the cathode area. The cathode of SOEC has higher inlet steam content than the anode chamber of a fuel cell. Compared with the tubular design that allows a separation between the oxidizing and reducing zones the edge sealing in a planar configuration is more difficult. Much effort is needed to address this issue.

Possible candidate sealing materials are listed as following: Silicate, borate, phosphate based glass systems and mica rex. Three general approaches are currently being pursued: rigid bonded sealing, compressive sealing, and compliant bonded sealing.

Reliability is also a major issue in cell seal development because during long periods of operation, and especially during thermal cycling, the thermal profile can vary considerably, creating a variety of stresses [46]. Rather than using traditional glass-based materials, which provide a rigid seal, PNNL researchers are working with mica materials to construct a more flexible, compressive seal. No one sealing technique will likely satisfy all stack designs and system applications. By comparison, the development efforts on compressive sealing have been more limited in scope but good progress has been achieved with hybrid mica seals. The concept needs to be tested on full-size components and test stacks to identify potential design and performance issues with scale-up.

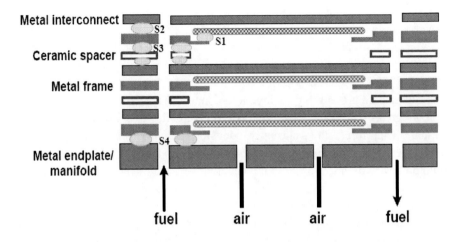

Figure 13. Schematic of seals typically found in a planar design SOFC stack [45].

Researchers are using a gasket approach, where mica materials are put between cell components and an external load is applied, making it similar to an "O" ring or a head gasket in a car. The mica-based seal is a more forgiving seal in terms of thermal expansion mismatch because it is made of parallel layers that de-bond from each other under high temperatures [47-48]. As a result, if a certain cell component expands significantly more than an adjacent component, the mica material can mechanically de-couple the two components, preventing the buildup of destructive stresses.

3.1.2. Interconnect

Interconnect is a critical component in SOEC. Irrespective of planar or tubular cell configuration, the role of interconnect is literally twofold; it provides electrical connection between anode of one individual cell to the cathode of neighboring one. It also acts as a physical barrier to protect the air electrode material from the reducing environment of cathode side, and it equally prevents the hydrogen electrode material from contacting with oxidizing atmosphere of the anode side [49-50]. The criteria for the interconnect materials are the most stringent of all cell components. In particular, the chemical potential gradient stemming from considerable oxygen partial pressure differences between the cathode and anode severe constraints on the choice of material for the interconnect.

The use of metallic interconnection between planar cells would result in a lower ohmic loss, improved resistance to thermal and mechanical shock, and reduced manufacturing costs. However, The choice of interconnection material is closely related to the choice of electrolyte, since the ionic resistivity of the electrolyte is temperature dependent, dropping by a magnitude of two between 800 and 900°C. Metallic interconnections would have to operate at lower temperatures than present day ceramic interconnect [51].

High growth rate and easy volatilization are the two main drawbacks of chromia. Therefore, the currently developed chromium-bearing metallic interconnects can only be applied for operating under 700°C. Future research should be directed on [52-53]: (1) the development of new interconnect materials with low thermal expansion and high electrical conductivity that can operate in both reducing and oxidizing atmospheres; (2) the development of new interconnect alloys from a fundamental understanding of the oxidation kinetics and oxide conductivity; (3) the development of compliant metallic interconnect designs in combination with novel and low cost stack design concepts; (4) the development of effective and low cost protective coating materials for metallic interconnects; (5) the investigation of novel coating approaches for metallic interconnects based on chemical, physical vapor deposition and thermal spray techniques; (6) reducing operating temperatures of SOEC to 800°C without compromising the power density.

3.1.3. Electrolyte Performance

The properties of the electrolyte have a major impact on electrolysis cell performance through its contribution to the ohmic internal resistance. An ideal SOEC electrolyte should have the following characteristics: high oxide ion conductivity (typically $> 1 \times 10^{-3}$ $S \cdot cm^{-1}$), low electronic conductivity, good thermal and chemical stability in relation to the reactant environment and the contacting electrode materials, closely matched thermal expansion coefficient (TEC) between the electrodes and contacting components, fully dense structure to maximize conductivity and minimize reactant cross-over, simple forming properties so that very thin layers can be fabricated as well as be low cost and environmentally benign.

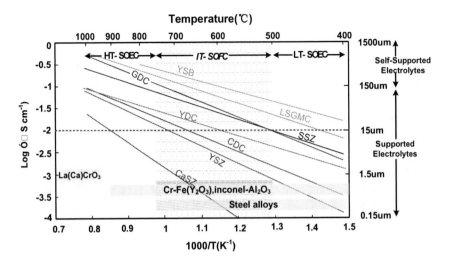

Figure 14. Variation of specific ionic conductivity of selected electrolytes with temperature [51].

Table 8. Comparison of electrolyte candidate materials

	Advantages	Disadvantages
YSZ	• Excellent stability in oxidizing and reducing environment • Excellent mechanical stability (3YSZ) • >40000 h of fuel cell operation possible • High quality raw materials available	• Low ionic conductivity (especially 3YSZ) • Incompatible with some cathodic materials
ScSZ	• Higher ionic conductivity • Excellent stability in oxidizing and reducing environment • Better long term stability than 8YSZ	• Scarce availability • High price of scandium
GCO	• Higher ionic conductivity • Good compatibility with cathode materials	• Poor mechanical stability • Electronic conductivity leading short circuit
LSGM	• Higher ionic conductivity • Good compatibility with cathode materials	• Ga evaporation at low P_{O2} • Incompatible with NiO • Poor mechanical stability

Figure 14 shows the variation of specific ionic conductivity of different solid oxide electrolytes with temperature [51]. These are representative values of conductivity for each of the electrolytes, actual values will depend on the microstructure, exact level of doping, fabrication and sintering processes. The oxide ion transport mechanism in ceramic electrolytes is thermally activated and the conductivity is a strong function of temperature. Ionic conductivity exhibits an Arrhenius-like dependence on temperature, which is the reason for the graph being plotted as the logarithm of conductivity vs. reciprocal of temperature. Advantages and disadvantages of possible electrolyte candidates for solid oxide electrolytes are listed in table 8 as below: Advanced materials (e.g., scandium doped zirconia, lanthanum gallate with strontium doping) are being developed and will be examined for the application in SOEC [55-57].

3.1.4. Anode Material

Anode materials must satisfy the following requirements [58]: mixed ionic-electronic conductivity, thermal and chemical stability at high temperatures in air and good chemical and thermo-mechanical compatibility with the electrolyte. Mixed ionic-electronic conductivity and a significant open porosity enable oxygen reduction not only on the surface but also in the entire volume of the electrode.

Most anodes are porous cermets (a composite of ceramic and metal), the microstructure of which is optimized to have a fully percolated metallic component which allows conduction of electrons through the structure, while optimizing the amount of active three phase boundary (TPB). The TPB is the interface at which the electronic and ionic conducting phases co-exist with the open pore containing hydrogen and steam as well as where reaction takes place in most cermets [59]. Optimization of the microstructure to maximize the effective TPB length is a major research effort in such cermets. However, the use of mixed ionic and electronic conducting (MIEC) ceramic materials is also being investigated so that the TPB can be extended from a 1-dimensional interface to a 2-dimensional area.

Perovskite-structured ceramic, Lanthanum strontium manganite (LSMC) has been the most frequently used material in SOEC. The disadvantage of LSMC is the thermal expansion and chemical incompatibility with YSZ electrolyte which will lead to a higher over-voltage related loss because of the growth of secondary phases such as $La_2Zr_2O_7$ and $SrZrO_3$ at the anode/electrolyte interface (see figure 15). The possible solution could be a Ce based interlayer in combination with mixed conducting cathode or new anodic material development for high temperature usage [61].

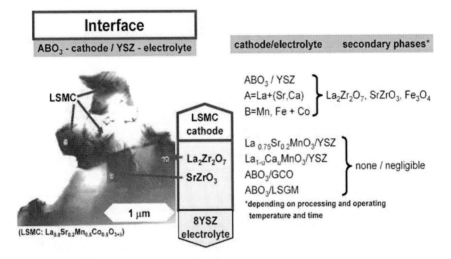

Figure 15. Formation of secondary phases at the interface [60].

3.1.5. Cathode Material

Porous Ni/YSZ cermet is currently the most common hydrogen electrode material for SOFC /SOEC applications, for its reasonable electro-catalytic activity, low cost, excellent chemically stability, appropriate thermal expansion coefficient [62]. Among SOEC components, the porous hydrogen electrode serves to provide electrochemical reaction sites for the decomposition of steam, allow the steam to be delivered and removed from the surface sites, and to provide a path for electrons to be transported from the interconnect to the electrolyte/anode reaction sites in SOEC stacks.

Researches on cathode material mainly focus on the agglomeration-degradation of Ni-YSZ after SOEC operation. Investigations of Ni/YSZ under SOEC operating conditions indicate that the microstructure optimization is of great importance. Moreover, the cell performance shows significant differences when operated in the different cell mode. The groups from Risø National Laboratory DTU reported that a short-term passivation occurs for the first few hundred hours, which could be attributed to the impurities [63, 64]. It was proposed that at least a part of these impurities originated from the applied albite glass sealing used in the setup for cell testing. However, their subsequent research showed that besides passivation, the main part of the long-term degradation affects from the sealing material of SOECs is caused by increasing losses in the hydrogen electrode. High current density can result in significant microstructural changes at the hydrogen electrode/electrolyte interface due to relocation of Ni. The mechanism behind the long-term degradation is not understood yet. Efforts need to optimize the microstructure for avoiding agglomeration [65].

3.1.6. Modular Design

Modular design of HTE system is a potential solution for meeting the different-scale demand of hydrogen production due to the limited size of single cells. Modular design arranges a group of uniform sized single cells together as a SOEC stack [66]. Then these SOEC stacks can be assembled to make a basic module. The hydrogen production capacity of each cell is directly related to the cell active area. Obviously the more the module, the larger the surface area results in higher hydrogen production capacity. The modular design makes itself to scale-up in size, which included increase in the size of module sub-assemblies and development of sub-systems required for system control operation. It also increases the system flexibility and the stack reliability. Moreover, the module can be standardized and controlled individually. If one individual module fails, it can be switched off and replaced.

In 1985, the concepts for integrate modular electrolysis units made up of series and parallel-connected tubular cells were presented and demonstrated by Doenitz and Erdle [67, 68]. Nowadays, modular design is also adopted for the Laboratory-Scale HTE system experiment in Idaho National Laboratory (INL), USA. Four sixty-cell stacks are combined into a module and each cell has an active area of 64 cm^2, providing a total active area of 15,360 cm^2 in a module. They are designed to operate in cross flow, with the steam / hydrogen gas mixture entering the inlet manifolds and exiting through the outlet manifold. To preclude the loss of an entire stack if a single cell fails, the four stacks are electrically interconnected at every fifth cell [69].

Currently, the technology challenges of SOEC stacks still remain in the areas of durability, thermal cycle, stack life and system integration. Also, the thermal and powder management of modular stacks are critical issues impacting the performance and cost of HTSE system. Moreover, when coupled with HTGR reactor, the safety and operation control of the large-scale system will arise to an important position.

3.1.7. System Management

SOEC stack is the key component of HTE. However, similar to the fuel cell system, HTE system also includes many subsystems besides SOEC stack itself, such as heat management subsystem, power management subsystem, steam management, gas transfers/purification subsystem, data acquisition and control subsystem and security subsystem [70].

Efficient heat management can improve the overall efficiency of HTE process by the reduction of energy consumption and the recover of some heat contained in the outlet products of the electrolysis [71]. The hydrogen production rate of HTE system can be changed flexible with power supply management. The HTE system is more complex comparing with SOFC systems. The main reason is that the steam is easy to be condensed, which will result in the crack of SOEC stacks and fluctuation of operation condition. Steam management subsystem can guarantee the effective use and accurate control of steam. The main function of gas transfers/purification subsystems is to control the composition of inlet gas, and to purify the product. Data acquisition and control subsystem can monitor and control the SOEC modules. In a HTE system, safety precautions are needed to be taken for handling high temperature hydrogen and oxygen effectively.

3.1.8. Preliminary Economic Analysis of HTE

As a promising way for massive production of hydrogen in the future, HTE must be cost effective in economy away from the technical advantages when compared with other methods, It is still very difficult to predict the hydrogen production cost precisely based on the current degree of HTE technological maturity. However, it can be concluded that SOEC used to host HTE process must withstand high temperatures while equally being reliable, economical and with a practical life cycle. Previous studies of CEA, France, showed that the hydrogen production cost is mainly linked to two factors: the hydrogen production plant investment and the electricity, heat consumption during electrolysis process [72, 73, and 74]. The low cost of electricity from nuclear power plant is an advantage of HTE coupled with HTGR power plant. The importance of the quantities of thermal energies (as steam) supplied by a nuclear reactor allows an extensive production of hydrogen on the same site. There is no doubt that this advantage influences certainly on investment and production costs. On the other hand, both hydrogen production plant investment and consumption of electricity, heat are related to SOEC system performance. e.g. higher hydrogen production efficiency, lower materials cost, better reality and durability. Also, it's a key issue to improve the performance of HTE process and the related equipments continuously for the economical improvement.

3.2. Coupling HTE with High Temperature Resource

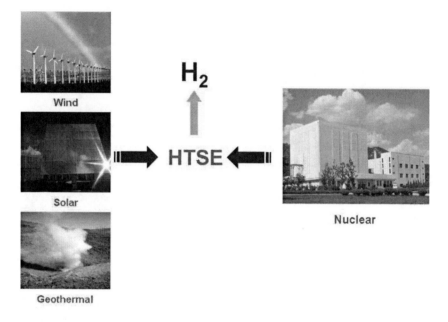

Figure 16. Heat source option for HTE [74].

As shown in figure 16, HTE processes are considered in combination with high-temperature heat source of nuclear or some developing renewable energy (including wind, solar or geothermal energy) recommended by European Hi2H2 project [75].

3.2.1. Nuclear Energy

Although the renewable energy has the characteristics of a sustainable future and plays an increasing role in energy supply frames, the common disadvantages are low energy density and they are difficult to keep stable supply due to the limitation of geography, climate conditions or other factors. Nuclear energy is clean energy and has been used effectively more than half a century. With the development of high temperature gas cooled reactor, it can supply heat and electricity to HTE system simultaneously, the efficiency of hydrogen production through heat to hydrogen directly can be significantly higher than that of conventional heat to electricity and then electricity to hydrogen process. Also, the process doesn't produce green house gases CO_2 and is economically competitive [76].

The coupling with the various processes of production of hydrogen in co-generation would permit to obtain higher efficiencies (see figure 16) [77]. With the modular aspect of the steam electrolysis, a nominal quantity of energy resulting from the reactor can be dedicated to the processes of electrolysis. The other part of the power will be distributed between the production of electricity and hydrogen according to the policy. For instance, the plant might produce electricity during the day and hydrogen at night, matching its electrical generation profile to the daily variation in demand. If the hydrogen can be produced economically, this scheme would compete favorably with existing grid energy storage schemes [78]. The total thermal power of the reactor will remain unchanged; it is only the ratio between electricity and hydrogen energy that can be adjusted. However, the integrated technology with advanced nuclear reactor is still a key point for HTE development. Because the nuclear power plants would not sell their kilowatt-hours to the general grid but to a specific site of hydrogen facilities. In this way, the nuclear-chemical system could be better optimized as a unified entity.

3.2.1.1. Coupling Pattern of HTE with HTGR

According to French researcher G. Rodriguez's study of several scenarios for large production of hydrogen by coupling a high temperature reactor with HTE, HTE process can produce hydrogen through three different operating conditions (see figure 17) [79]: thermo-neutral (isothermal), allo-thermal (endothermal) and auto-thermal (exothermal). The evolution from one operating condition to another one depends on the value of the current density and the energy balance at the level of the electrolyzer. In the studied case, the allothermal condition will be situated in a range of current density between 0 and 0.98 Acm^{-2}. At the value of 0.99 Acm^{-2}, it is the thermoneutral condition. Above this value, the hydrogen will be produced according to the autothermal condition. Thermo-neutral point is occurring at thermal equilibrium condition when the inlet and outlet gas temperature are equal. The heat is mainly due to anodic and cathodic over-voltages as well as ohmic losses by Joule effects in the electrolytic membrane.

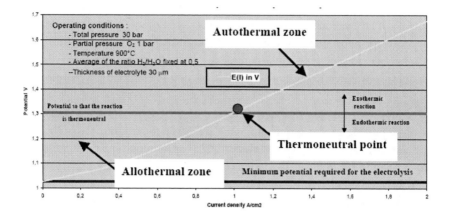

Figure 17. Three operating conditions of HTE Process.

Comparison of the three operating conditions is given in the following table 9 [80]. It is clear from table 9 that the allothermal mode represents the best results. Under the same operating temperature, the autothermal mode is less efficient (50.8%). The second is the thermoneutral condition with a yield equaling to 51.3%. The best is the allothermal mode with a value of 54.6%. The thermoneutral mode is demanding the less total amount of energy (608 MW), the other operating modes are almost of the same value.

Table 9. Comparison of different operating modes

	Autothermal	Thermoneutral	Allothermal
Temperature	900℃	900℃	900℃
Production of hydrogen	327MW	312MW	353MW
Yield	50.8%	51.3%	54.6%
Power provided by helium	0	39.8MW	45.3MW
Thermal energy to add to the electrolyser	0	0	42.2MW
Total amount of energy required	643.7MW	608MW	646.4MW

An independent study conducted by U.S. Argonne National laboratory (see table 10) showed that the capital cost of a plant with $1 kgH^2/h$ hydrogen production scale through HTE is the same as that through low temperature electrolysis(conventional electrolysis, LTE) though the component costs are different from each other[81]. But for the thermal power of the nuclear reactor, which is consumed both for electricity production and for heating the steam to the desired electrolysis temperature, the total thermal energy extracted from the nuclear reactor can be represented as the sum of the thermal energy from the nuclear reactor used for electrolysis and the heat energy, 156.4kW are needed for HTE, whereas it takes 200kW for LTE on the assumption η_{el}=35% (The calculation is listed as the following), thus the overall energy required is reducing and the process efficiency is improving 21.8%. This is one of the primary advantages of HTE.

$$if \quad \eta_{el} = 35\%$$

$$then \quad Q_H = Q_e / \eta_{el} + Q_h = 51.02/35\% + 10.66 = 156.43kW$$

$$Q_L = Q_e / \eta_{el} + Q_h = 70/35\% + 0 = 200kW$$

$$and \quad \Delta\eta = (Q_L - Q_H)/Q_L = (200-156.4)/200 = 21.8\%$$

The opposite consideration is that H_2 is storable and should be stored for commercialization. Discontinuous operation of the electrolytic facilities is possibility to be considered, although continuous operation seems to be more adequate from the electrochemical point of view [82-83]. As already noted in relation to the generation IV initiative, scenarios of nuclear H_2 production would need very safe and economically competitive reactors. In the Next Generation Nuclear Plant (NGNP), the high temperature thermal energy produced by the nuclear reactor is transmitted to the hydrogen production plant through an intermediate heat exchanger (IHX), a heat transfer loop, and at least one process heat exchanger (PHX). Technically, the challenges of creating such a system are many, and much research thus far has been on the development of enabling technologies including materials identification and property measurements, component design, measurement of physical data to fill knowledge gaps, and the development and testing of integrated system models [84].

Table 10. Comparison of Low Temperature and High Temperature Electrolysis (1kgH^2/h)

	Capital Cost Comparison		Power Use Comparison	
	High-T	Low-T	High-T	Low-T
Housing/Forecourt dispenser (NEPA Compliant)	$18,000	$12,000		
Electric Power Transformer	$10,000	$14,000	2.50	3.50
Water System-manifolding	$500	$1,250	0.10	0.10
Steam system	$8,000		0.00	0.00
Electric Power Inventor/Conditioning	$30,820	$45,932	1.81	2.70
Electrolytic Cells($/kW)	$123,282	$143,199	36.26	54.04
Hydrogen Compression(ambient-100psi)	$27,000	$20,000	1.05	1.05
Hydrogen compression(100-5000psi)	$78,678	$78,678	3.60	3.60
Hydrogen dispensing (5000psi)	$17,984	$17,984	0.50	0.50
800kW-Natural Gas Compression (50- 3600 psi)	$29,974	$29,974	2.09	2.09
800kW- Hythane dispensing(3600psi)	$11,990	$11,990	0.50	0.5
Oxygen system manifolding	$20,000	$2,500	0.70	0.15
Controls/Auxiallry	$25,076	$25,585	1.20	0.8
Moisture dryers	$2,150		0.00	0.00
Cooling fan	$875	$1,250	0.71	0.98
Profit	$60,671	$60,657		
Total	$465,000	$465,000		
Power(kW electric)			51.02	70.00
Power(kW Q@1050K)			10.66	0.00

3.2.1.2. Intermediate Heat Exchanger (IHE)

One of the key issues in the nuclear hydrogen production systems is a Intermediate heat exchanger (IHE) [85]. Hydrogen production using a high-temperature reactor requires the efficient transfer of the high-temperature heat from the primary reactor coolant to the HTE process. It is anticipated that the heat source and the hydrogen process plant will be coupled through an intermediate heat transfer loop that provides the desired degree of isolation between nuclear and hydrogen production plants. The high-temperature intermediate loop poses unique challenges in materials, heat exchanger design, safety, and supporting system designs. Hydrogen or process heat plants involve high temperatures and often involve corrosive or toxic chemical species associated with the intermediate-loop heat exchanger in the hydrogen plant [86].

The helium gas flows in one side of the heat exchanger and high temperature steam flows in the other side of the heat exchanger. The fabrication of the heat exchanger includes a machining of the flow path, a coating and ion beam mixing, and a diffusion bonding of the heat transfer plate. The materials used for IHE require excellent mechanical properties at an elevated temperature as well as a high corrosion resistance in a high humidity environment [87]. A ceramic heat exchanger with a strong corrosion resistance has difficulties for its manufacturing. Ceramic material can be used at a higher design temperature exceeding $1000\,°C$. However, its machining and bonding are serious problems for its high temperature applications. These problems can be considered to be eliminated for the SiC coated and ion beam mixed metallic heat exchanger. Also a hybrid design and surface modification technology could be implemented for the integrated system in order to increase its design pressure and its corrosion.

3.2.2.2. Nuclear Plant/Hydrogen Plant Safety

Unlike electricity, high temperature heat can only be transported limited distances. thus, the nuclear reactor and hydrogen plants will be close to each other. A critical issue is to understand potential incremental safety risks to the nuclear reactor from the associated hydrogen plant to assure nuclear plant safety. The challenges are mainly focused on the following four points [84]. 1) Because the safety philosophy for the hydrogen plant is different from the safety philosophy for nuclear power plants, this difference must be recognized and understood when considering safety issues to a nuclear reactor from coupled hydrogen plants; 2) Accidental releases of hydrogen, deuterium and tritium from a nuclear plant are able to penetrate through metals. It is of great importance to study the possibility of tritium from the reactor core to infiltrate into the secondary loop. 3) Many chemical or hydrogen plants under accident conditions can produce heavy ground-hugging gases such as oxygen, corrosive gases, and toxic gases that can have major off-site consequences because of the ease of transport from the plant to off-site locations. Oxygen presents a special concern because most proposed nuclear hydrogen processes convert water into hydrogen and oxygen; thus, oxygen is the primary byproduct. These types of potential accidents must be carefully assessed; 4) the potential consequences of the failure of the intermediate heat transport loop that moves heat from the reactor to the hydrogen plant must be carefully assessed.

During the operation, three tiers of events may happen in an integrated nuclear and hydrogen plant: (1)accidents or equipments failures at hydrogen plant or IHE loop lead nuclear reactor core damage and significant release of radioactive materials from nuclear reactor, (2) leaks of chemicals from hydrogen plant into environment. Such events may lead injury to operators, (3) unscheduled plant occurrences (unplanned shutdown, minor equipments failures, etc.) disturb normal operation of plants. Of the tree tiers, the first tier events are most concerned. The related authorities must be convinced that the co-location and connection of a hydrogen plant to nuclear reactor pose no significant increased hazards the nuclear plant.

Quantitative risk analysis (QRA, also called probability risk assessment) is a useful tool for understanding the incremental risk related to the hydrogen plant. In QRA, possible hazards are first identified, and then are analyzed for fresuency and impact. The relationships between individual events and overall impacts of the events may be complex and the probabilities and consequences of individual events must be assessed to arrive at an overall understanding of risk.

Figure 18. Integration of HTTR with HTE process and IS cycle [77].

In this area an initial work carried out in INL is the evaluation of the minimum separation distance between the nuclear plant and the hydrogen plant [78, 87]. The events analyzed in this study were hydrogen explosions and the release of chemical clouds from a co-located Sulfur-Iodine hydrogen plant. The goal of the study was to determine the minimum distance needed between the nuclear plant and hydrogen plant such that the incremental probability of causing nuclear reactor core demage due to hydrogen explosion or a chemical release was no greater than $1.0E^{-6}$ event/year. The conclusions from this study were that a minimum separation distance of 110m was needed between the nuclear plant and the hydrogen plant in the default configuration for hydrogen explosion involving up to 100 kg H_2 (as shown in figure 18).

3.2.2. Renewable Energy

3.2.2.1. Geothermal Energy

A feasibility study exploring the use of geothermal energy in hydrogen production is underway. It is possible to use a thermal energy to supply heat for high temperature electrolysis and thereby substitute a part of the relatively expensive electricity needed. Geothermal fluid is used to heat fresh water up to 200°C steam. The steam is further heated to 900°C by utilizing heat produced within the electrolyser. The electrical power of this process is reduced from 4.6 kWh/Nm^3 of hydrogen for conventional process to 3.24 kWh/Nm^3 for HTE process implying electrical energy reduction of 29.5%. The geothermal energy needed in the process is 0.5 kWh/Nm^3. Price of geothermal energy is approximately 8~10% of electrical energy and therefore a substantial reduction of production cost of hydrogen can be achieved this way. It will be shown that using HTE process with geothermal steam reduces the production cost by approximately 19%. However, there are a variety of technical and regulatory challenges preventing the more widespread use of geothermal energy [88].

- Many of the best potential resources are located in remote or rural areas, where are difficult to develop projects;
- Leasing and sitting processes can take long periods and be fraught with uncertainty;
- Although costs have decreased in recent years, exploration and drilling for power production remain expensive;
- The productivity of geothermal wells may decline over time

3.2.2.2. Solar Energy and Wind Power

Hirsch and Steinfeld demonstrated in 2004 that the linkage of steam-electrolysis to a solar energy system would be a very encouraging progress which can improve system scalability. Some European countries also consider integrating HTE system with wind power [89].

The main challenge for the renewable energy industry is that the costs of power generation by most renewable energy are too high and the market accessibility is limited. At present, the average cost of power generated by renewable energy is much higher than that generated by traditional energy. For example, in China, the cost of a small hydropower is 20% higher than that of coal power, biomass (biogas) is 50% higher and wind power is 70% higher. The cost of photo voltaic power is 10 to 17 times higher. The high costs prevent the expansion of the market share of renewable energy and the limited market share creates obstacles to creating incentives for cost reduction. Thus HTE integrated with nuclear energy can be used in the near term while the integration with renewable energy can be considered potential pathways for the long-term.

4. HISTORY OF HTE STUDIES

The concept of High-temperature electrolysis production of hydrogen from steam was first investigated in the 1980s by the German Dornier-system GmbH and Lurgi GmbH in cooperation with Robert Bosch GmbH in the process called "High Operating Temperature Electrolysis, HOT ELLY" [67,68,90]. Their R&Ds were mainly focused on first generation electrolyte supported tubular SOEC and relative assembly technologies as well as demonstration tests using a pilot plant. SOEC had been practically developed by Dornier-system GmbH, and self-supporting electrolysis tubes were fabricated by connecting solid-oxide in series. Basic hydrogen production data on HTE was obtained with a 10 solid oxide cell (10mm length each) tube by Doenitz et al. The mainly experimental parameters were listed in table 11. Hydrogen was yielded at the maximum rate of 6.8 Nl/h with an efficiency of 92% at 970°C and a power of 21.7W. Based on the preliminary experimental results, Doenitz integrated an HTE module consisting 10 modules (including 1000 electrolysis cells) were assembled into a stack which can produce hydrogen at the maximum rate of 0.6Nm^3/h. However, the stability in the long term operation, the reliability of the electrolysis tube and the module against thermal cycles were not published. This project was closed down around 1990 due to the low fossil fuel prices at that time.

Table 11. Main parameters for HTE by Doenitz

Electrolyte	YSZ (Y_2O_3 8~12mol%)	Current density	370mA/cm^2
Cathode	Ni-YSZ	Electrolytic power	21.7W
Anode	Sr doped LaMnO$_3$	Operation temperature	970°C

Westinghouse Electric. Co. also made its efforts to carry out HTE experiment through a tubular single cell in the 1980s [91]. The electrolyte and the electrodes layers were formed on a closed-one-end porous support tube made of calcia-stabilized ZrO_2(CSZ), whose porosity was around 35%(12mm in inner diameter, 1~1.5 mm in thickness and 1000mm in length). The main experimental parameters were listed in table 12. Hydrogen was produced at a maximum rate of 17.6Nl/h at 1000°C and the current density of 400mA/cm^2.

Table 12. Main parameters for HTE by Westinghouse Electric. Co.

Electrolyte	CSZ	Current density	400mA/cm^2
Cathode	Ni-YSZ	Electrolytic power	39.3W
Anode	Sr doped LaMnO$_3$	Operation temperature	1000°C

In Japan, Japan Atomic Energy Research Institute (JAERI) initiated its HTE program in 1995 [92, 93]. The goal of this work is to obtain design data for the HTE module and to accumulate experineces of operational procedures on HTE. Laboratory-scale experiments were carried out with practical tubular and planar SOEC, respectively. The main experimental parameters were listed in table 13. Hydrogen was produced at the maximum density of 44Ncm^3/cm^{2}h at 950°C by using a 12-cell tube electrolysis cell. The experimental correlation of the hydrogen production density was derived in relation with the electrolysis temperature and the applied current density. Also improving a steam flow structure can result in the increase of hydrogen production density due to sufficient steam supply to the cathode. Tests were also carried out with planar electrolysis cells. Hydrogen could be produced continuously at the maximum density of 38Ncm^3/cm^{2}h at 850°C which showed almost the same performance of the electrolysis tube obtained at 950°C. However, both the tubular cell and planar cell couldn't keep their integrity in serveral thermal cycle and Efficiencies achieved were still on a very low level without advanced high temperature heat source [94].

It was not until November 29, 2004 that Researchers at the U.S. Department of Energy's Idaho National Laboratory (INL) and Ceramate, Inc. of Salt Lake City announced a breakthrough development in hydrogen production from nuclear energy [95-98]. They achieved the highest-known production rate of hydrogen by HTE (about 45%~52%) while the laboratory current efficiency is almost 100%. This development is viewed as a crucial first step toward large-scale production of hydrogen from water, rather than fossil fuels and a milestone in the hydrogen energy research field, which was hoped to help the United States advance the President George W. Bush's Hydrogen Fuel Initiative and the goal of a clean hydrogen economy. Department of Energy (DOE) of the United States hoped that INL can commercially produce hydrogen production by HTE before 2017 [99].

Table 13. Main parameters for HTE by JAERI

Tube SOEC	Electrolyte	YSZ (Y_2O_3 8mol%)
	Cathode	Ni-YSZ
	Anode	$LaCoO_3$
	Operation temperature	950 °C
Planar SOEC	Electrolyte	YSZ (Y_2O_3 8mol%)
	Cathode	Ni-YSZ
	Anode	$LaMnO_3$
	Operation temperature	850°C

Single button cell tests, multi-cell stack, as well as multi-stack testing have been conducted in INL. SOEC test loop for stack and modular stack testing is shown schematically in figure 19. Argon is being used as a carrier gas to control the steam-hydrogen ratio being fed to the button cell or stack in the furnace. The dewpoint temperatures of the feed stream and the exiting stream are directly measured to determine the amount of hydrogen produced by the button cell or stack. For safety reasons, the oxygen is diluted with air before being exhausted. The dewpoint measurements correlate well with expected hydrogen production rates based on measurements of electrical current supplied to the cell. Stack testing used 100cm^2 cells (64cm^2 active areas) supplied by Ceramatec and ranged from 10 cell short stacks to 240 cell modules. Tests were conducted either in a bench-scale test apparatus or in a newly developed 5 kW integrated laboratory scale (ILS) test facility. The ILS facility has been designed for an ultimate nominal hydrogen production rate of 14.1kW based on lower heating value (LHV, equal to 120MJ/kg for hydrogen), or 4735 Normal L/h. The initial ILS facility single module implementation was designed for 5 kW hydrogen productions. Neither heat recuperation nor hydrogen recycle were incorporated. In the facility, four sixty-cell stacks are combined into a module as shown in figure 20(a). Each cell has an active area of 64cm^2 per cell, providing a total active area of 15360cm^2 in a module. They are designed to operate in cross flow, with the steam/hydrogen gas mixture entering the inlet manifolds on the right and left sides, and exiting through the outlet manifold visible. Airflow enters through an air inlet manifold and exits through the front and back open faces directly into the hot zone enclosure. Each pair of stacks is called a half module, shown in figure 20(b). To preclude the loss of an entire stack if a single cell fails, the four stacks are electrically interconnected at every fifth cell. The facility is now being expanded to handle 3 solid modules for a total of 720 64 cm^2 cells [100].

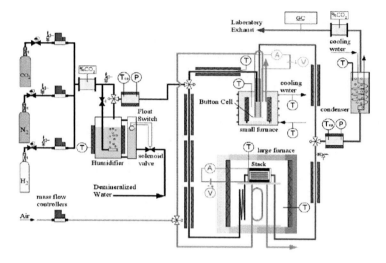

Figure 19. Testing Apparatus for Button Cells and Stacks [100].

The critical parameters for a 300MW commercial electrolysis plant have been determined based on these experimental results. A preliminary design for modular electrolyzer units, to be used in conjunction with a high temperature heat source, has also been developed. The major achievements associated with HTE in INL are summarized in Table 14.

In addition, GE company in the USA carried out a research project on High Performance Flexible Reversible Solid Oxide Cell supported in part by DOE award DE-FC36-04GO14351 from 2005 [101]. The objectives of this project are mainly focused on: 1) Demonstrate a single modular stack that can be operated under dual modes. 2) Provide materials set, electrode microstructure, and technological gap assessment for future work. A joint university of nevada, Las Vegas and Argonne national laboratory project was initiated from 2005 to understand and optimize surfaces, interfaces and layer properties using a multifaceted approach to address the challenges for HTE [102]: 1) Elucidate the surface and bulk reaction mechanisms at the O_2 electrode; 2) Determine the oxidation state of the nickel and the structure of the electrode/electrolyte interface during cell operation at H_2/H_2O electrode.

Figure 20(a). ILS 4-stack module and (b) One-half of ILS module [100].

Table 14. Major achievement associated with HTE in INL from 2004 to 2008

Year	Achievements of HTE
2004	6 Cell, 64cm^2, H$_2$:30NL/h
2005	10 Cell, 64cm^2, H$_2$:60NL/h
2006	25 Cell, H$_2$:160NL/h,1000h
2007	Syntrolysis, Recommend for one of the 100 most technologically significant products of 2007
2008	Module, 3×240 cells, 64cm^2, 4735NL/h

In the European Union, the Highly Efficient, High Temperature, Hydrogen Production by Water Electrolysis (Hi2H2) project had been started by the participation in activities within the 6th Framework Programme of the EU from the end of 2004 [103]. The coordinator included European Institute for Energy Research, Swiss Federal Laboratories for Materials Testingand Research,

Deutsches Zentrum für Luft- und Raumfahrt (German Aerospace Center) and Risø National Laboratory. The goal of the project are [74]: 1) Demonstrate the feasibility of using planar SOC technology for high temperature electrolysis; 2) Analyze the limitations and degradation mechanisms of SOFC cells used in HTE mode; 3) Develop new materials for corrosion resistant and high performance HTE. The long-term plan is to collaborate with DOE to demonstrate a centralized plant coupled to a Gen4 nuclear reactor or collaborate with Japan to demonstrate a plant coupled to their HTR which is in operation. Risø National Laboratory had put its efforts on building a test setup to investigate the state of art SOFCs stack operated as SOECs with the focus on the kinetics and durability. Results obtained within the project have demonstrated that hydrogen can be produced through water electrolysis with higher efficiencies, greater than 90%. SOFC can be operated in the electrolysis mode with low operation voltages in the range of 1 to 1.3V. A world record-breaking performance was reached at 950°C, where steam electrolysis was operated at a current density of 3.6 A/cm^2 and a cell voltage of only 1.48 V. Cells are thus suitable for reversible operation. Also the economy of HTE for hydrogen production is estimated with the cost of fossil fuels. H_2 can be produced at attractive production costs. The hydrogen production cost was found to be 30US$/barrel equivalent crude oil using the HHV and given the assumption of electricity price 3.6US$/GJ. The CO production cost was found to be 40 US$/barrel equivalent crude oil using the HHV and given the same assumption [104].

The RELHY project (Innovative Solid Oxide Electrolyser Stacks for Efficient and Reliable hydrogen production) was funded under 7th FWP (Seventh Framework Programme) [105]. The RELHY project targets the development of novel or improved, low cost materials (and the associated manufacturing process) for their integration in efficient and durable components for the next generation of electrolysers based on SOEC. The Participants includes: Helion Sas, European Institute for Energy Research, Danmarks Tekniske University, Topsoe Fuel Cell, Imperial College of Science, Technology and Medicine and Energy Research Center of the Netherlands.

The University of Iceland and Icelandic New Energy are currently preparing a campaign to work towards promoting HTE for hydrogen production [106]. To this end, cooperation with CEA of France is under way in association with the University of Grenoble. The objective is to incorporate graduate projects and exchange of students and specialists between France and Iceland in order to realize the application of HTE technology.

In Asia, China, Japan and Korea, HTE programs have also been started with the goal to bring nuclear hydrogen production (thermal-chemical water splitting cycle and HTE) to the energy market in recent 3 years [107-109]. The Projects will treat the pertinent aspects of material development, HTGR fuel technology, nuclear waste management, coupling to hydrogen production technologies.

5. HTE DEVELOPMENT AT INET IN CHINA

China is taking a very active approach for developing hydrogen energy technology including production, storage and application technology of hydrogen. In the "Tenth Five –Year Plan (2001-2005)", funding for EV & H/FC related programs added up to 40% of total energy research budget. It focuses mainly on basic and technical aspects of hydrogen energy and related demonstration projects such as fuel cell city bus, refueling station and Hydrogen Park. An example is successful service provided by hydrogen powered automobiles during Beijing Olympic Games in Aug. 2008 [110].

Since 1970's, high temperature gas cooled reactor (HTGR) technology has been developed in Institute of Nuclear and New Energy Technology (INET),Tsinghua University. 10 MWth test reactor (HTR-10) with spherical fuel elements was constructed in 2000 and now is in operation. This test reactor can be utilized to develop a modular high temperature gas-cooled reactor design, as well as to establish an experimental base for nuclear process heat application. A number of safety related experiments have been conducted in HTR-10. R&D on direct cycle helium turbine technology is being carried out. Coupling a helium turbine system to the existing 10 MWth test reactor is foreseen. The construction of industrial scale demonstration plant of modular HTGR (HTR-PM) is one of the National Major S&T Special Projects. The construction of 200MWe HTR-PM will be finished around 2013. The related design, construction and R&D work has been started [111].

Hydrogen production by nuclear energy is a promising way for industrial scale production comparing with other developing hydrogen production methods. Among all nuclear reactors, HTGR is most suitable for nuclear hydrogen process due to its potential of high efficiency power generation cycle and providing high temperature process heat, therefore, R&D on nuclear hydrogen, as a part of the HTGR-PM project, has started in INET, Tsinghua University. S-I thermo-chemical cycle for splitting water and high temperature steam electrolysis (HTE) are selected as potential process for nuclear hydrogen production. Since 2005, INET has conducted preliminary study on IS process and HTE process. The laboratory of nuclear hydrogen with facilities for process studies has been established. HTR-10 constructed in INET will provide a real nuclear facility for future research and development of nuclear hydrogen technology.

Development of nuclear hydrogen in china will undergo the following 3 phases:

- Phase one (2008-2013): Process verification of NH process and bench scale test
- Phase two (2014-2020): R&D on coupling technology with reactor, nuclear hydrogen safety and pilot scale test
- Phase three (after 2020): Commercialization of nuclear hydrogen production

The current R&D activities on nuclear hydrogen produsiton in INET are focusing on realizing the target of phase one.

5.1. HTE DEVELOPMENT SCHEDULE OF INET

The R&D on HTE development are divided into four stages as shown in figure 21: 1) From 2008 to 2009, construction of HTE test facilities and process verification, 2) From 2010 to 2013, bench-scaled experimental study with hydrogen production yield of 60L/h, 3) From 2014 to 2020, the design of pilot plant and the pilot scale test with hydrogen production yield of $5Nm^3/h$ as well as R&D on the coupling technology with HTGR, 4) Commercial demonstration after 2020 [25].

Currently, the research activities mainly focus on: 1) Demonstrate the feasibility of using planar SOEC technology for high temperature electrolysis; 2) Develop new materials with corrosion resistant and high performance HTE; 3) Analyze the degradation mechanisms of SOEC cells used in HTE mode; 4) HTE cell and stack optimization; 5) System design studies to support cycle life assessment and cost analysis for HTE plants.

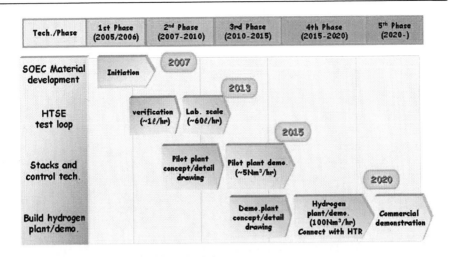

Figure 21. High temperature electrolysis R&D plan in INET.

5.2. R&D ON HTE

The research and development of HTE technology was initiated at INET in 2005. Currently, two testing systems, one is for HTE cell online testing and another is for high temperature electrochemical performance evaluation of SOEC components have been designed and constructed as shown in figure 22 (a) and (b) . In addition, the research on novel anode materials has obtained excellent results. The theoretical analysis of hydrogen production efficiency of HTE coupled with HTGR has been carried out. The obtained results have been discussed in Section 2.2.1.

(1) Study on Conventional Planar LSM-SOEC System

The lab-scale hydrogen production on conventional LSM-SOEC system was investigated. The electrolyte layer made of YSZ (containing 8% mol% of Y_2O_3) was sandwiched between the porous cathode (Ni/YSZ) and anode layer (LSM). Under the same current density, the overpotential voltage decreases and the electrolysis performance improves with increasing temperature. When the input voltage is 1.0V and the temperature is 850°C, when the input voltage is 1.018v, the hydrogen production density is 2.23 ml/min • cm^2. When the voltage increases to 1.318V, the hydrogen production density increases to 5.15 ml/min • cm^2 correspondingly.

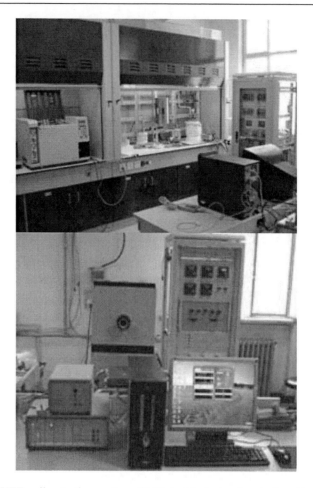

Figure 22. (a) HTE online testing system. (b) Material electrochemical performance evaluation system.

Table 15 shows the area–specific resistance (ASR) of LSM electrodes under SOEC and SOFC modes, respectively. From table 15, it could be seen that the polarization resistance of a Ni-YSZ/YSZ/LSM cell stack was only 0.76 Ω.cm^2 and had no obvious change when running in SOFC mode but it increased about 4 times, which was 3.7Ω.cm^2 when operating in SOEC mode. The study of Risø national laboratory also showed the similar results. Therefore, we can see that although the high temperature electrolysis is essentially a reverse process of SOFC in principle, the conventional materials of SOFC are not suitable for operation in SOEC mode.

Table 15. ASR of LSM-SOEC under different operating modes at 950℃

ASR ($\Omega.cm^2$) \ Mode \ Anode	SOFC	SOEC
LSM	0.76	3.7

(2) Development of Novel Anode Materials with Low ASR [115-117]

The feasibility of novel conductive membrane $Ba_xSr_{1-x}Co_{0.8}Fe_{0.2}O_{3-\delta}$ used as oxygen electrode of SOEC was studied from several aspects of the Goldschmidt tolerance factor, critical radius, lattice free volume, average bond energy, variable valence capability of B elements, as shown in figure 23(a) and (b). A strategy for the systematic selection of oxygen electrode material was explored and the optimum combination of the A site(x=0.5), that is $Ba_{0.5}Sr_{0.5}Co_{0.8}Fe_{0.2}O_{3-\delta}$(BSCF) was selected out.

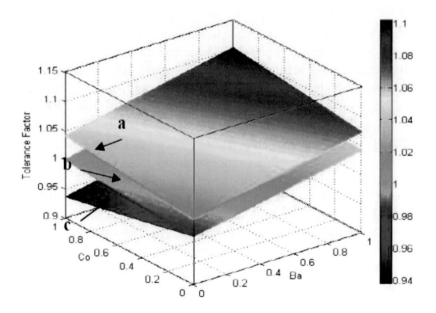

(a) $Ba_{1-x}Sr_xCo_{1-y}Fe_yO_3$ (b) $Ba_{1-x}Sr_xCo_{1-y}Fe_yO_{2.5}$ (c) $Ba_{1-x}Sr_xCo_{1-y}Fe_yO_{2+y}$

Figure 23.(a) Calculated results of tolerance factor of $Ba_{1-x}Sr_xCo_{1-y}Fe_yO_{3-\delta}$

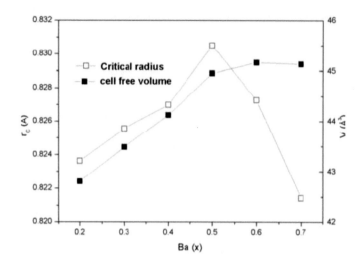

Figure 23.(b) Critical radius and cell free volume of $Ba_xSr_{1-x}Co_{0.8}Fe_{e0.2}O_{3-\delta}$ with various Ba contents.

ASR is one of the most important characteristic parameters in measuring the electrolysis ability of SOEC for hydrogen production. It can directly represent the level of electrochemical performance of BSCF oxygen electrodes. The lower the ASR value is, the higher the performance of anode electrode is, which means the stronger oxygen permeability and conductivity of BSCF. Compared with other oxygen electrode materials, ASR data of the electrode BSCF/YSZ are $0.66 \, \Omega \, cm^2$ at 750°C, $0.27 \, \Omega \, cm^2$ at 800°C, and only $0.077 \, \Omega \, cm^2$ at 850°C, remarkably lower than the common used oxygen electrode materials LSM as well as the current focused materials LSC and LSCF, as shown in figure 24(a).

Figure 24(b) is hydrogen production rate of SOEC prepared by BSCF and LSM anodes at various electrolysis voltages (half cell with cathode and electrolyte of YSZ/Ni-YSZ is kindly supplied by Shanghai Institute of Ceramics, Chinese Academy of Sciences) under the same current density of 300mA, respectively. From figure 24(b), it can be seen that the hydrogen production rate of both BSCF/YSZ/Ni-YSZ cell and LSM/YSZ/Ni-YSZ cell increase with the increase of electrolysis voltage. When the voltage is up to 1.4V, the hydrogen production rate of BSCF-cell is $147.2 \, mL \cdot cm^{-2}h^{-1}$, about three times as that of LSM-cell (about $49.8 \, mL \cdot cm^{-2}h^{-1}$), which indicates that BSCF could be a potential candidate for the application of SOEC anode.

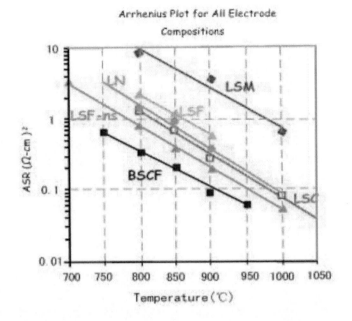

Figure 24.(a) ASR of BSCF in comparison with other oxygen electrodes.

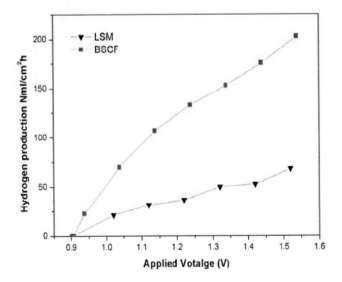

Figure 24.(b) Hydrogen production rate of SOEC prepared by BSCF and LSM anodes at various electrolysis voltages, respectively.

(3) Microstructure Control of Cathode [118-121]

Previous analysis indicated that the coarsening and the oxidation of nickel particles as well as the diffusion of steam were the limited step in the whole electrolysis reaction. As gas permeability and electrical conductivity of SOEC cathodes are strongly dependent on the cathode microstructure, the reasonable control of the microstructure is crucial for the optimization of the electrochemical performance of the cathode.

In order to enhance performance, the study of hydrogen electrodes focuses on two aspects. 1) Prepare the materials of the functional layer with the microstructure as fine as possible via a novel in-situ coating combustion method to extend the length of TPB, 2) Screen novel pore formers to get suitable porosity, pore shape and distribution of supporting cathodes for SOEC application.

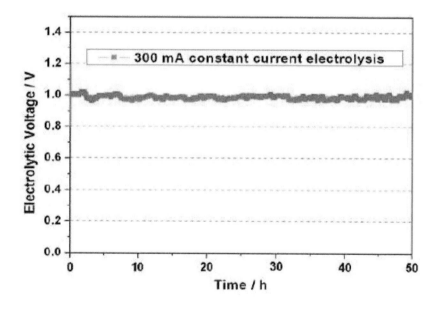

Figure 25. Stability of the single button cell.

Nano-sized NiO powder on submicron-sized YSZ particles of functional layer was prepared via in-situ coating combustion method. XRD and FESEM analysis showed that the products were well crystallized with NiO coating on YSZ particles. The optimized addition ratio of $CO(NH_2)_2$ to $Ni(NO_3)_2$ was 2:1. A SOEC single cell made from NiO-YSZ with the molar ratio of 2:1 composite powder exhibited better performance than the other samples with the electrolytic voltage of 0.98V and showed excellent durability (zero degradation) under an electrolytic current density of 0.33 A/cm^2, an input stream composition of 80%H$_2$O+20%H$_2$ and a temperature of 900°C for 50 h (see figure 25).

Four different pore formers, including polymethyl methacrylate (PMMA), potato starch, ammonium oxalate, ammonium carbonate, were considered for the optimization. Their influences on the amount of porosity and on the pore shape and distribution as well as the effect on the electronic conductivity were analyzed. The results showed that PMMA was the most promising pore former, which had high porosity and uniform pore size distribution. The optimum weight percent concentration was 10%, correspondingly, porosity was 45% and electronic conductivity was 6726S·cm^{-1}, which was suitable for supporting cathodes for SOEC application.

(4) Design of the Stack and the HTE Stack Online Testing System

Figure 26 is the design of the modular HTE stack online testing system. The whole system mainly consists of three parts: measure & control part, gas loop part, and hydrogen monitoring part. The whole system is being under construction and the construction will be finished in 2009.

Figure 27 is the design of the planar stack. The short stack is designed to be composed of three planar individual cells with a hydrogen production rate of 1L/h. Table 16 indicates some of the cell configuration details that have been adopted for this conceptual design. The stacks are planned to be assembled in July, 2009. After that, it will be tested in HTE stack online testing system.

Table 16. Cell configuration and technological specifications

Cell Configuration					Specifications	
	Composition	Thickness	Size	Porosity/ Density	Temp.	850°C
Electrolyte	YSZ	10-20μm	6.5×6.5	D>95%	Input Steam	>70%
Anode	LSM	30-50μm	5×5	P>20%		
Cathode	Ni-YSZ	1000μm	6.5×6.5	P>35%	Efficiency	>90%
Seal	Glass-Ceramic	3000μm		D>95%	ASR-C	<1 Ω. cm^2
Bipolar plate	Ferrite		6.5×6.5		ASR-U	<1.5 Ω. cm^2
Channel Width	1.0mm	Bipolar Thickness		3.0mm	Degradation rate	<0.05% /5h
Channel Width	0.5mm	Ridge Width		1.0mm		

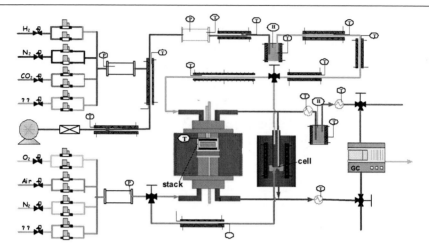

Figure 26. Design of the modular HTE testing loop.

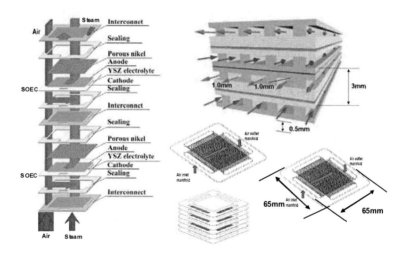

Figure 27. Design of the stack.

6. CONCLUSION AND DISCUSSIONS

The researches of HTE in several countries are giving very promising results, which highlight the fact that the HTE process can be a suitable technology for massive production of hydrogen in the next decades. The hydrogen production plant can be coupled with advanced nuclear reactor HTGR. The integrated system has several distinct advantages including:

- No greenhouse gas CO_2 emissions:
- High efficiency. High temperature reduces the electricity needed for electrolysis and speed the mass transport in the process, so very high efficiency could be achieved .The thermal-to- hydrogen conversion efficiency may be as high as 45~59%;
- High Flexibility. HTE plants could serve an important balancing function for HTGR power plant, i.e. supplying both electricity and hydrogen. Also, Modular SOEC stack design has flexibility for different scale of hydrogen demand.

Although HTE technology has been well developed in the past years, there is a great gap for the commercialization. For closing the gap, it is necessary to get more financial support for R&D on HTE technology and to establish broader international cooperation.

REFERENCE

[1]   World Energy Outlook, International Energy Agency, *OECD/IEA Publications,* Second Edition, 2002.

[2]   Adger, N., Kelly, P.M., Ninh, N.H., (2001). *Living with Environmental Change.* London, Routledge.

[3]   Hydrogen Strategy Group of the Federal Ministry of Economics and Labour (2005). *Strategy Report on Research Needs in the Field of Hydrogen Energy Technology.* Berlin, Federal Ministry of Economics and Labour.

[4]   Johnston B., Mayo C. M., Khare A. (2005). Hydrogen: the energy source for the 21st century. *Technovation,* 6,569-585.

[5]   Jiang, Z.M. (2008).Reflection on Energy Issues in China. *Journal of Shanghai Jiaotong University,* 3,345-359.

[6]   Hirsch D., Steinfeld A. (2004), Solar hydrogen production by thermal decomposition of natural gas using a vortex-flow reactor, *Hydrogen Energy,* 29,47-55.

[7]   Yi B. L. (2004). Fuel Cell: Principle, Technology and Application (in Chinese). Beijing: Chemical Industry Press.

[8]   Charles W., Forsberg K., (2001). Hydrogen production as a major nuclear energy application, DE-AC05-00OR22725. US.

[9]   Barreto L., Makihira A., Riahi K. (2003).The hydrogen economy in the 21st century: a sustainable development scenario, *International Journal of Hydrogen Energy,* 28,267-284.

[10] Carte J. D.,Myers D., Kumar R., Recent Progress on the Development of TuffCell, a Metal-Supported SOFC/SOEC, January 22-27 2006, The 30th International Conference & Exposition on Advanced Ceramics and Composites, Cocoa Beach Florida

[11] Virkar A., Tao G.,The State of the Art of SOFC: Power Generation and Reversible Cells, 2006. *Third International Workshop on Fuel Cells –* WICaC.

[12] Bollinger R. B., Aaron T M, Low Cost Hydrogen Production Platform, 2002,U.S. DOE Hydrogen Program Review.

[13] Brown, L.C., G.E. Besenbruch, J.E. Funk, A.C. Marshall, P.S. Pickard, S.K. Showalter , High Efficiency Generation of Hydrogen Fuels Using Nuclear Energy, Presentation at U.S. Department of Energy Hydrogen Fuel Cells and Hydrogen Review, Nuclear Energy Research Initiative,2002

[14] World Nuclear Association, Information and Issue Briefs:Hydrogen Economy, <http://www.world-nuclear.org/ info/inf70.htm>, June 2003.

[15] Gobina E., Hydrogen as a Chemical Constituent and as an Energy Source, February 2003. *Norwalk. Business Communications* Company (BCC) Report C-219R.

[16] Joel M., Pham A., Aceves S., (2003).A natural gas-assisted steam electrolyzer for high-efficiency production of hydrogen, *International Journal of Hydrogen Energy,* 28,483-490.

[17] International Atomic Energy Agency. Hydrogen as an energy carrier and its production by nuclear power. *IAEA-TECDOC*-1085, 1999:101-130.

[18] Yurum, Y, Hydrogen production methods. *Fuel and Energy Abstrct.* 1996, 37:344.

[19] Odgen J. Review of small stationary reformers for hydrogen production, IEA-H$_2$ report, 2002, < www.eere.energy.gov/afdc/pdfs/31948.pdf>.

[20] Yildiz B., Kazimi M., (2006).Efficiency of hydrogen production systems using alternative nuclear energy technologies, *International Journal of Hydrogen Energy,* 31,77-92.

[21] Spacil H. S., Tedmon C. S. (1969).Electrochemical dissociation of water vapor in solid oxide electrolyte cells I.Thermodynamics and Cell Characteristics. *Journal of the Electrochemical Society,* 116, 1618-1626.

[22] Bolthrunis C. O., Allen D. , Goff K.,et al. Investigation of the impact of temperature on hydrogen production cost from advanced water splitting technologies.Proceedings of the 4th International Topical Meeting on High Temperature Reactor Technology,HTR2008,September 28-October 1, 2008, Washington, DC,US.

[23] Tinoco R. R., Mansilla C., Bouallou C.. Technoeconomic study of the high temperature steam electrolysis process coupled with a sodium nuclear reactor. .2008,17 th World Hydrogen Energy Conference, Brisbane, Australia.

[24] Fujiwara S., Kasai S., Yamauchi H. , et al. (2008).Hydrogen production by high temperature electrolysis with nuclear reactor. *Progress in Nuclear Energy*, 50,422-426.

[25] Yu B.,Zhang W.Q.,Chen J.,et al. (2008).Advance on highly efficient hydrogen production by high temperature steam electrolysis, Science in China Series B: *Chemistry,* 51,289-304.

[26] O'Brien J., Stoots C., Herring J., Lessing P. , Hartvigsen J. , Elangovan S. (2005),Performance Measurements of Solid-Oxide Electrolysis Cells for Hydrogen Production, *Journal of Fuel Cell Science and Technology,* 2,156-163.

[27] Zhang W. Q., Yu B., Chen J. (2006).Progress on the Anode Materials for Solid Oxide Fuel Cells (SOFC) and Its Application for Hydrogen Production through High Temperature Steam Electrolysis, *Progress in Chemistry* (in Chinese),18,:832-840.

[28] Ni M., Leung M.K.H.,et al. (2006).A modeling study on concentration overpotentials of a reversible solid oxide fuel cell. *Journal of Power Sources,* 163,460–466.

[29] Sridhar K.R., Vaniman B.T. (1997).Oxygen production on Mars using solid oxide electrolysis. *Solid State Ionics,* 93,321-328.

[30] Wang W.S., Huang Y., Jung S.,et al. (2006).A Comparison of LSM, LSF, and LSCo for Solid Oxide Electrolyzer Anodes. *Journal of The Electrochemical Society,* 153,A2066-A2070.

[31] Zhang W. Q., Yu B., Chen J, Xu J.M. (2008).Hydrogen Production Through Solid Oxide Electrolysis at Elevated Temperatures, *Progress in Chemistry* (in Chinese),20,778-787.

[32] Liu M. Y., Yu B, Xu J M, Chen J. (2008). Thermodynamic analysis of the efficiency of high-temperature steam electrolysis system for hydrogen production. *J. Power Source, 177,*493-499.

[33] Kazimi Y., Efficiency of hydrogen production systems using alternative nuclear energy technologies. *GA report on SI process,* 2003.US.

[34] M.Y. Liu, B. Yu, J. Chen, J.M. Xu, (2008).Thermodynamic Analysis of the Efficiency of High Temperature Steam Electrolysis (HTSE) System for Hydrogen Production, *J. Power Source, 177,*493-499.

[35] Shin Y., Park W.,Chang J.,Park J. (2007). Evaluation of the high temperature electrolysis of steam to produce hydrogen. *Int. J. Hydrogen. Energy,* 32, 1486-1491.

[36] Wu Z.X., Zhang. Z.Y., (2004). Advanced Nuclear System and High-Temperature Gas-Cooled Reactor, Beijing ,Tsinghua University Press.

[37] Youngjoon S., Wonseok P., Jonghwa C., Jongkuen P., (2007).Evaluation of the high temperature electrolysis of steam to produce hydrogen, *International Journal of Hydrogen Energy,* 32,1486-1491.

[38] Herring J, Lessing P, O'Brien J, et al. Hydrogen Production through High-Temperature Electrolysis in a Solid Oxide Cell, *OECD/NEA Workshop on Nuclear Production of Hydrogen,* 2-3 October 2003.ANL, US.

[39] Herring J.S., O'Brien J.E., Stoots CM, Lessing PA, Anderson RP, Hartvigsen JJ, et al. Hydrogen production through high-temperature electrolysis using nuclear power. April 25–29 2004, AIChE spring national meeting, New Orleans.

[40] Momma A. , Kato T.,Kaga Y. et al, (1997).Polarization behavior of high temperature solid oxide electrolysis cells (SOEC), *Journal of the ceramic Society of Japani,* 105,369-373.

[41] Herring J., Lessing P., O'Brien J.Hydrogen Production through High-Temperature Electrolysis in a Solid Oxide Cell , Second Information Exchange Meeting on Nuclear Production of Hydrogen, 2 and 3 October 2003, Argonne National Laboratory, Illinois, US.

[42] Herring J, O'Brien J, Lessing P, Nuclear Energy and the Hydrogen Economy, MIT, September 23, 2003, US.

[43] Herring J., O'Brien J., Stoots C.,et al. FY 2003 Hydrogen, Fuel Cells, and Infrastructure Technologies Progress Report, 2003,US.

[44] Hauch A.,Jensen S.H.,Mogensen M.,Ni/YSZ Electrodes in Slid Oxide Electrolyser Cells,2006,Proceedings of the 26th Risø International Symposium on Materials Science: Solid State Electrochemistry, DK.

[45] Stevenson J., SOFC Seals: Materials Status, SECA Core Technology Program-SOFC Seal Meeting July 8, 2003, Sandia National Laboratory, Albuquerque, NM

[46] Steele B.C.H., (2000).Materials for IT-SOFC stacks 35 years R&D: the inevitability of gradualness? *Solid State Ionics,* 134, 3-20.

[47] Lessing P.A.,(2007),A review of sealing technologies applicable to solid oxide electrolysis cells, *J. Mater Sci.* 42,3465-3476.

[48] Weil K.S., (2006).The State-of-the-Art in Sealing Technology for Solid Oxide Fuel Cells, JOM, 58,37-44.

[49] Molenda J., (2006).High-temperature solid-oxide fuel cells new trends in materials research, *Materials Science*-Poland, 24, 1-6.

[50] Haile S.M., (2003). Fuel cell materials and components, *Acta Materialia,* 51, 5981-6000.

[51] Brett D.J.L., Atkinson A., Brandon N. P.,et al. (2008).Intermediate temperature solid oxide fuel cells, *Chem. Soc. Rev.,*37,1568-1578.

[52] Zhu W.Z., Deevi S.C. (2003).Development of interconnect materials for solid oxide fuel cells, Materials Science and Engineering, A348, 227-243.

[53] Singh P., Minh N. Q.,(2004). *Solid Oxide Fuel Cells:* Technology Status, *International Journal of Applied Ceramic Technology,* 1,5-15.

[54] Liang M D , YU B.,.Wen M F, Xu J M, Zhai Y C. (2008). The fabrication technique of YSZ electrolyte film. *Progress in chemistry.* 20. 1222-1232.

[55] Osada N.,Uchida H.,Watanabe M., (2006).Polarization Behavior of SDC Cathode with Highly Dispersed Ni Catalysts for Solid Oxide Electrolysis Cells, *Journal of The Electrochemical Society*, 153,A816-A820.

[56] Lessing P.A., (2007).Materials for hydrogen generation via water electrolysis, *J. Mater. Sci.* 42, 3477–3487.

[57] Ni M., Leung M.K.H., Leung D.Y.C. (2008), Technological development of hydrogen production by solid oxide electrolyzer cell (SOEC). *International journal of hydrogen energy,* 33, 2337-2354.

[58] Uchida H., Osada N. ,Watanabe M., (2004).High-Performance Electrode for Steam Electrolysis Mixed Conducting Ceria-Based Cathode with Highly-Dispersed Ni Electrocatalysts, *Electrochemical and Solid-State Letters,* 12, A500-A502.

[59] Singhal S.C., Kendall K., (2003).High-temperature solid oxide fuel cells: fundamentals, design and applications. Elsevier, US.

[60] Hauch A., Ebbesen S. D., Jensen S.H., Mogensen M.(2008).Highly efficient high temperature electrolysis. *Journal of Materials Chemistry,* 18,2331-2340.

[61] Tao Y., Nishino H., Ashidate S.,et al. (2009).Polarization properties of $La_{0.6}Sr_{0.4}Co_{0.2}Fe_{0.8}O_3$-based double layer-type oxygen electrodes for reversible SOFCs. *Electrochimica Acta,* 54,3309–3315.

[62] Jiang S. P., Chan S.H. (2004).A review of anode materials development in solid oxide fuel cells, *journal of materials science,* 39,4405-4439.

[63] Hauch A., Jensen S. H., Ramousse S.,et al, (2006).Performance and Durability of Solid Oxide Electrolysis Cells, *Journal of The Electrochemical Society,* 153,A1741-A1747.

[64] Hauch A.,Jensen S. H. , Bilde-Sørensen J. B.,et al, (2007).Silica Segregation in the Ni/YSZ Electrode, *Journal of The Electrochemical Society,* 154,A619-A626.

[65] Hauch A., Ebbesen S. D., Jensen S. H.,et al, (2008).Solid Oxide Electrolysis Cells: Microstructure and Degradation of the Ni/Yttria-Stabilized Zirconia Electrode, *Journal of The Electrochemical Society,* 155,B1184-B1193.

[66] Richards M., Shenoy A., Schultz K.,et al,The Modular Helium Reactor for Hydrogen Production, 15th Pacific Basin Nuclear Conference, 2006, Australia.

[67] Döenitz W., Eedle E. (1985).High-temperature electrolysis of water vapor status of development and perspective for application. *International Journal of Hydrogen Energy,* 10,291-295.

[68] Döenitz W.,Dietrich G.,Erdle E.,et al.(1988).Electrochemical high temperature technology for hydrogen production or direct electricity generation. *International Journal of Hydrogen Energy,* 13,283-287.

[69] Stoots C. M., O'Brien J. E., Herring J. S. ,et al, Idaho national laboratory experimental research in high temperature electrolysis for hydrogen and syngas production, *Proceedings of the 4th International Topical Meeting on High Temperature Reactor Technology* HTR2008, September 28-October 1, 2008, Washington DC ,US.

[70] Vielstich W., Lamm A., Gasteiger H. A.,(2003).Handbook of Fuel Cells: *Fundamentals, Technology, Applications,* UK, Wiley.

[71] Mansilla C., Sigurvinsson J., Bontemps A. ,et al, (2007).Heat management for hydrogen production by high temperature steam electrolysis, *Energy,* 32,,423-430.

[72] Ni M., Leung M.K.H., Leung D.Y.C., (2006).A modeling study on concentration overpotentials of a reversible solid oxide fuel cell, *Journal of Power Sources,* 163 ,460-466.

[73] Winkler W., Koeppen J. (1996).Design and operation of interconnectors for solid oxide fuel cell stacks, *Journal of Power Sources,* 61,201-204.

[74] RiveraTinoco R., Mansilla C., Bouallou C. hydrogen production from high temperature electrolysis: economic impact of the interactions of the electrolyser investment, lifespan and performance, 2008, 17 th World Hydrogen Energy Conference, Brisbane, Australia.

[75] Hi2H2 project, Highly efficient, high temperature, hydrogen production by water electrolysis, <http://www.hi2h2.com/>, [accessed 20.04.09].

[76] Brisse A., Hauch A., Schiller G.,et al, highly efficient, high temperature, hydrogen production by water electrolysis (HI2H2),17th world hydrogen energy conference,,15-19 June,2008, Queensland, Australia.

[77] Utgikar V., Thiesen T. (2006). Life cycle assessment of high temperature electrolysis for hydrogen production via nuclear energy. *International Journal of Hydrogen Energy,* 31,939-944.

[78] Smith C., Beck S., Galyean W., Separation Requirements for a Hydrogen Production Plant and High Temperature Nuclear Reactor, INL-EXT-05-00137, October 2006.

[79] Herring J. S., Lessing P., O'Brien J. E. Hydrogen Production through High-Temperature Electrolysis in a Solid Oxide Cell, Second Information Exchange Meeting on Nuclear Production of Hydrogen, 2 - 3October 2003. Argonne National Laboratory, Illinois, US.

[80] Rodriguez, G. Pinteaux, T. Studies and design of several scenarios for large production of hydrogen by coupling a high temperature reactor with steam electrolysers, French CEA report, 2003.

[81] WERKOFF F. , MARECHAL A,. Techno Economic Study on the Production of Hydrogen by High-temperature Steam Electrolysis. French CEA report, 2003.

[82] Center for Transportation Research, ANL/ESD/TM-163, June. 2001. Argonne National Laboratory, US. <http://www.transportation. anl.gov/pdfs/TA/153.pdf>.

[83] Mireia P., Martínez-Val J. M., Montes M. J., (2006).Safety issues of nuclear production of hydrogen,Energy Conversion and Management,47,732-2739.

[84] Baindur S., safety issues in nuclear hydrogen production with the very high temperature reactor (VHTR), Proceedings of the 29th Canadian Nuclear Society Annual Conference, June 1-4, 2008, Toronto, Canada.

[85] Sherman S. R., (2007). Nuclear plant/hydrogen plant safety: issues and approaches. embedded Topical Meeting: Safety and Technology of Nuclear Hydrogen Production, Control, and Manage, Boston, Massachusetts. US.

[86] Chang H. O., Kim E. S., Design Option of Heat Exchanger for the Next Generation Nuclear Plant, Proceedings of the 4th International Topical Meeting on High Temperature Reactor Technology, HTR2008, September 28-October 1, 2008, Washington, DC US

[87] Forsberg C. W., Gorensek M. , Safety Related Physical Phenomena for Coupled High-Temperature Reactors and Hydrogen Production Facilities , Proceedings of the 4th International Topical Meeting on High Temperature Reactor Technology,HTR2008,September 28-October 1, 2008, Washington, DC US

[88] McDonald C. F. (1996). Compact buffer zone plate-fin IHX—The key component for high-temperature nuclear process heat realization with advanced MHR, *Applied Thermal Engineering,* 16,3-32.

[89] Bragi A, Thorsteinn I, 2006 Report of Sigfussion Science Institute, University of Iceland.

[90] Jensen S. H.,Larsena P. H., Mogensen M., (2007).Hydrogen and synthetic fuel production from renewable energy sources. *International Journal of Hydrogen Energy*, 15, 3253-3257.

[91] Döenitz W., Schmidberger R. (1982).Concepts and design for scaling up high-temperature water vapor electrolysis. *International Journal of Hydrogen Energy*, 7,321-330.

[92] Maskalick N. J. (1986).High temperature electrolysis cell performance characterization. *International Journal of Hydrogen Energy*, 11,563-570.

[93] Hino R, Miyamoto Y. Hydrogen production by high temperature electrolysis of steam. Proceedings of a technical committee meeting, October 19–20, 1992 International Atomic Energy Agency, Oarai, Japan, IAEA-TECDOC-761: 120-4.

[94] Hino R., Aita H., Sekita K., Haga K., Iwata T. (1997). Study on hydrogen production by high temperature electrolysis of steam. Japan Atomic Energy Research Institute, JAERI-Research 97-064.

[95] Hino R., Haga K., Aita H.,et al, (2004). 38. R&D on hydrogen production by high-temperature electrolysis of steam, *Nuclear Engineering and Design*, 233,363-375.

[96] Matthew L. Wald, Hydrogen Production Method Could Bolster Fuel Supplies, New York Times November 28, 2004.< http://www.nytimes.com/ 2004/11/28/ politics/28hydrogen.html?_r=1>.

[97] O'Brien J. E., Stoots C.M., Herring J. S.,et al, (2005).Performance Measurements of Solid-Oxide Electrolysis Cells for Hydrogen Production, *Journal of Fuel Cell Science and Technology*, 2,156-163.

[98] O'Brien J. E., Stoots C.M., Herring J. S.,et al, (2006).Hydrogen Production Performance of a 10-Cell Planar Solid-Oxide Electrolysis Stack, *Journal of Fuel Cell Science and Technology*, 3,213-219.

[99] Herring J. S., O'Brien J. E., Stoots C.M., et al, (2007). Progress in high-temperature electrolysis for hydrogen production using planar SOFC technology, *International Journal of Hydrogen Energy*, 32, 440-450.

[100] David Henderson, Nuclear Hydrogen Initiative Overview, May 24, 2004, US. <www. hydrogen.energy. gov/ pdfs/review04/ 3_nhi_overview_ henderson.pdf>

[101] Stoots C.M., O'Brien J.E., Herring J. S.,et al. Idaho national laboratory experimental research in high temperature electrolysis for hydrogen and syngas production. Proceedings of the 4th International Topical Meeting on

High Temperature Reactor Technology, HTR2008, September 28-October 1, 2008, Washington, DC US.

[102] Mansilla C.,Sigurvinsson J.,Bontemps A., (2007). Maréchal A., Werkoff F., Heat management for hydrogen production by high temperature steam electrolysis, *Energy*, 32,423-430.

[103] Shin Y., Park W., Chang J., Park J. (2007). Evaluation of the high temperature electrolysis of steam to produce hydrogen. *Int. J. Hydrogen Energy*, 32, 1486-1491.

[104] Amitava R., Watsona S. , Infield D., (2006), Comparison of electrical energy efficiency of atmospheric and high-pressure electrolysers, *International Journal of Hydrogen Energy*, 31,1964-1979.

[105] Jensen S. H., Larsen P. H., Mogensen M. (2007) .Hydrogen and synthetic fuel production from renewable energy sources. *International Journal of Hydrogen Energy*, 32, 3253-3257.

[106] Relhy project,Innovative Solid Oxide Electrolyser Stacks for Efficient and Reliable Hydrogen Production. < http://www.relhy.eu/>.

[107] Arnason B., Sigfusson T. I. Application of geothermal energy to hydrogen production and storage. 2nd German Hydrogen Congress, Essen; February 2003. Published in Proceedings. <http://theochem.org/bragastofa/ CD/ essen.pdf>.

[108] Lee S., Kang K. H., Kim J. M.,et al. (2008). Fabrication and characterization of Cu/YSZ cermet high-temperature electrolysis cathode material prepared by high-energy ball-milling method I. 900°C-sintered, *Journal of Alloys and Compounds*, 448,363-367.

[109] Lee S., Kim J. M., Hong H. S.,et al. (2009).Fabrication and characterization of Cu/YSZ cermet high-temperature electrolysis cathode material prepared by high-energy ball-milling method II. 700°C-sintered, *Journal of Alloys and Compounds*, 467, 614-621.

[110] Fan C. G., Iida T., Murakami T.,et al (2008).investigation on the power generation and electrolysis behavior of Ni-YSZ/YSZ/LSM cell in reformate Fuel. *Journal of Fuel Cell Science and Technology*, 5:031202.1-031202.5.

[111] Zhang Ping, YU B., Present status and future plan of nuclear hydrogen production program in INET. 2007, The third information meeting of nuclear hydrogen production, Oct.4-7, Oarai, Japan,. (Nuclear Hydrogen Production, OECD).

[112] Shi D.H. (2006). Chinese hydrogen update,5 IPHE steering committee meeting, 28-29 March, Canada.

[113] Yu B., Zhang W.Q., Chen J., Xu J.M., Status and Research of Highly Efficient Hydrogen Production through High Temperature Steam

Electrolysis at INET,The Fourth International Hydrogen Forum, August 4-8, changsha Changsha, China.

[114] YU B., Zhang W. Q., Xu J. M. Research Status of High Temperature Steam Electrolysis at INET. International hydrogen forum 2008, August 4-8, changsha, China, 2008 (Accepted by *International Journal of Hydrogen Energy*).

[115] Yu B., Zhang W. Q., Xu J. M., et al (2008). Microstructural characterization and electrochemical properties of $Ba_{0.5}Sr_{0.5}Co_{0.8}Fe_{0.2}O_{3-\delta}$ and its application for anode of SOEC. *International Journal of Hydrogen Energy*, 33, 6873-6877.

[116] Zhang W. Q., Yu B. Chen J., Xu J. M.(2007).Electrochemical performance of $Ba_{0.5}Sr_{0.5}Co_{0.8}Fe_{0.2}O_{3-\delta}$ synthesized by a novel citric acid-nitrate combustion method for SOEC. The 16th international conference on solid state ionics. Shanghai.

[117] YU B., Zhang W. Q., Xu J. M. Electrochemical characterization of BSCF oxygen electrode for SOEC and SOFC. 17th World Hydrogen Energy Conference, Brisbane, Queensland, Australia, June 15-19, 2008.

[118] Liang M. D., Yu B., Xu J.M., (2009) .Preparation of LSM–YSZ composite powder for anode of solid oxide electrolysis cell and its activation mechanism. *Journal of Power Sources,* 190, 341–345.

[119] Liang M. D., YU B., Wen M. F.. Preparation of NiO-YSZ composite powders by a new situ-combustion method. International hydrogen forum 2008, August 4-8, changsha, China, (Accepted by *International Journal of Hydrogen Energy*).

[120] Liu M. Y., YU B., Xu J. M. Physical property and microstructure analysis of hydrogen electrode support layer of solid oxide electrolysis cell for hydrogen production. International hydrogen forum 2008, August 4-8, changsha, China (Accepted by International Journal of Hydrogen Energy).

[121] Liu M. Y., YU B., Xu J. M. Microstructural Characterization and Electrochemical Properties of Solid Oxide Electrolysis Cells for Hydrogen Production, 17th World Hydrogen Energy Conference, Brisbane, June 15-19, 2008Queensland, Australia.

INDEX